Nanotechnology

Nanotechnology

Rahul Rao

ICON

Published in the UK and USA in 2026 by
Icon Books Ltd, Omnibus Business Centre,
39–41 North Road, London N7 9DP
email: info@iconbooks.com
www.iconbooks.com

ISBN: 978-183773-238-8
eBook: 978-183773-237-1

Typeset by SJmagic DESIGN SERVICES, India.

Printed and bound in the UK.

Appointed GPSR EU Representative:
Easy Access System Europe Oü, 16879218
Address: Mustamäe tee 50, 10621, Tallinn, Estonia
Contact Details: gpsr.requests@easproject.com, +358 40 500 3575

For my parents and sister

CONTENTS

1 Zooming in   1

2 The world's tiniest assembly guide   13

3 Racing to the bottom   35

4 A fantastic voyage   57

5 Electrons at the nano gates   75

6 Going small to save the planet   101

7 Nano nightmares   123

8 Nano dreams   143

9 Zooming out   167

Further reading   175

Acknowledgements   178

Index   179

ZOOMING IN

1

What is nanotechnology? We might start answering that question with an example of what nanotechnology could theoretically do: take humans to outer space.

Today, rockets are our only practical means of reaching space. We need a rocket's spectacular power to overcome the formidable strength of Earth's gravity trying to pull us down. But rockets are unwieldy and costly. They rely on colossal tanks of bulky fuel, and every additional kilogram of payload we cram on board means additional fuel we must purchase to propel that kilogram into orbit. Today's cheapest rockets start at over £1,000 per kilogram. A trip to space is an event, not a practical part of everyday life.

But neither were skyscrapers practical parts of everyday life until the invention of the lift gave humans a machine to fend off gravity. Likewise, scientists, engineers and writers like Arthur C. Clarke have long dreamt of a lift on a far grander scale – an elevator to space.

Astronauts could board a space elevator at the ground and rise to orbit without the pyrotechnics of a rocket launch. In the other direction, a space elevator could offer a comfortable descent back to Earth without the dramatics of high-speed re-entry. A space elevator could eventually make a trip to space as routine as a train across the Channel.

It's a simple and elegant solution – sadly, too simple, and too elegant.

If a space elevator is to stay stable, its upper level must always remain above the same point on Earth's surface. Passengers will need to alight at the altitude where satellites today move just quickly enough to match Earth's rotation. This is known as geostationary orbit. Where the International Space Station orbits at about 400 kilometres above sea level, and most cosmonauts have never ventured more than 1,000 kilometres out, geostationary orbit lies at a far more rarefied 36,000 kilometres high – nearly the length of a full circle about Earth's equator. If the space elevator's cabin rose at the top speed of a London-to-Paris train, a voyage from sea level to geostationary orbit would last almost five full Earth days.

Our elevator will have to stretch further still – it's actually the elevator's centre of mass that will sit in geostationary orbit. That means its cable will need to stretch higher than any of the lift's riders will experience, out to a counterweight at an even greater altitude. Space elevator proposals would have the counterweight at a vertigo-inducing 100,000 kilometres up – more than a quarter of the distance to the Moon.

Building a space elevator will almost certainly be the largest construction project in human history. The greatest part of that expense is likely to be constructing the cable that will form the space elevator's backbone. If we want to build one, we'll need to forge a cable that's not only 100,000 kilometres long, but a cable that can also support itself at scales no human structure has ever done.

The key will be finding the correct material. We can start by looking at a material's breaking length: how tall a column we can build in Earth's gravity before the material buckles under its own weight. Designing a space elevator will be far more complex than simply stacking a material on top of itself like this, but breaking length gives us a useful first impression. Gravity weakens as we ascend the cable, but we'll still want a material with a breaking length of at least thousands of kilometres.

Concrete has a breaking length of less than half a kilometre. Steel fares better, but even the strongest known steel has a breaking length of only 30 kilometres or so. Spider silk, famously stronger than steel, reaches its limit at around 100 kilometres. Carbon-fibre can do even better but still buckles at about 400 kilometres.

Clearly, standard Earth building materials won't do for a space elevator, but where might we find a material that will support something so very big? Counterintuitively, many would-be space elevator architects suggest that we look for something very small: a carbon nanotube.

You won't find carbon nanotubes in any Earth lift. We've only been able to make them at all since the 1990s. Yet there is every reason to believe that we will want to make them in far larger quantities than we've got so far. Carbon nanotubes have the highest tensile strength of almost any known material; they can withstand stretching that's hundreds of times more severe than concrete or steel or carbon fibres can. Such mundane materials' breaking lengths pale in comparison to the carbon nanotube's, which may eclipse 4,000 kilometres. That places the space elevator back on the table.

What separates a carbon nanotube from a strand of spider silk, or a filament of carbon fibre, is its size. A single carbon nanotube is 100,000 times slimmer than a single human hair. Look at a cross-section of a carbon nanotube, and you can see the individual atoms that form its structure. Carbon nanotubes are so small that they don't belong to our world so much as to a parallel world of very, very small things: a nano world too small for most of us to see.

Entering the nano realm

We derive 'nano' from a Greek word for 'dwarf' or 'small person'.* Today, 'nano' has become a byword for 'small'. Your mobile may hold a nano SIM card; perhaps you remember the now-defunct iPod Nano. These are examples of manufacturers slapping a 'nano' label on products they deem too small for 'micro' to fit. The nano SIM is

---

* In fact, modern Greek uses the word 'nanos' for both dwarf stars and Tolkien's dwarves.

an unfathomable colossus next to the nano realm, which works on far, far smaller sizes. A carbon nanotube is about one nanometre across – one billionth of a metre.

How small, then, is a billionth of a metre? Perhaps the easiest way to find out would be to take after the titular character from the 1957 film *The Incredible Shrinking Man* and visit the nano realm ourselves.*

The average human being is slightly less than two metres tall. We are far, far too large to meet the nano realm face-to-face. So, imagine that we can shrink ourselves to, say, one-tenth of our original size. Shrink once, and we will be just shy of twenty centimetres tall, about the size of a modest laptop computer's screen. A twenty-centimetre-tall human is half the size of the average newborn, but we are nowhere near the nano realm.

Shrink once more, and we are now about two centimetres tall. We're still visible to the average human, but that full-sized human might easily lose us in a coat pocket. The average human foetus has already grown past our size by the end of its first trimester. We are about the size of an old iPod Nano.

Shrink again, down to just two millimetres, smaller than a nano SIM. We could hide behind a grain of rice or peek through the eye of a sewing needle. We might see

---

* Fortunately, our journey will be more pleasant than that of poor Scott, who starts shrinking after he is exposed to a noxious pesticide. As he shrinks away from human scales, he loses his job, must move into a dollhouse, is stalked by his cat and is eventually left for dead by his wife to fight a spider in his own cellar.

grains of pollen as large as party balloons and grains of sand and salt and sugar as large as pieces of furniture. We could comfortably ride on the back of an ant. From our newfound mount, we might look around and see many zoos' worth of life-forms that evaded our sight until we became this small: amoebas and paramecia swimming in water, mites that live in dust or human hair, even the very largest bacteria.

Shrink again, and we become the size of a single grain of sand. It's a sign of a changing landscape that we must now switch to less familiar units. We are now 200 micrometres tall; a micrometre, or micron, is one-millionth of a metre. At 200 micrometres, even the head of a pin might seem as large as a house. If a human hair is caught on this pin, that single strand of hair might seem as thick as a large pipe.

The naked human eye can resolve objects down to around 100 micrometres. If we shrink beyond this, we will well and truly leave the familiar human world behind. We're still far from the nano world.

Shrink again. At twenty micrometres tall, we are ourselves about the size of a single human skin cell. Regular-sized humans can only see us with the help of microscopes. We'll see cells all around us. If we look at the hair we just saw, we might be able to see the cells that comprise it, lined up inside its strand like bricks in a wall. We'll be able to see similar cells deep within each and every animal and plant, not to mention the cells of protists and bacteria that emerge into view from every corner.

Shrink again, and we plunge ever deeper into a world teeming with once-occluded life. All about us we may find cells of every shape and variety: white and red blood cells, yeast cells, liver cells, sperm cells. We are now the size of an *E. coli* bacterium, two micrometres large. We can now peer into a cell and see its organelles, or parts, such as the mitochondria that power it or its ribosomes that build proteins. If we look into a plant cell, we might get a close-up look at its individual chloroplasts, the organelles that catch the Sun's light and photosynthesise it into energy.

We are now passing the blurry threshold into the nano realm.

A cell's organelles can let us glimpse what the nano realm is capable of doing. They're performing tasks that don't sound out of place in a factory – generating energy, processing instructions, assembling proteins – yet each organelle is so small that its feedstock and spare parts are individual molecules.

Shrink again. We are now tinier than all but the very smallest cells. If we were to look into a cell's nucleus, we might spot twirling vines: the telltale double helix of DNA. We are now the size of a virus, and at last, we can measure our size in nanometres (nm) – we are now around 200 nm, or 200 billionths of a metre, tall. We are now smaller than the shortest wavelengths of visible light, making us all but invisible to even the best light-based microscopes.

Shrink again, one last time.

We are now 20 nm tall, small enough to swing on those DNA strands as if they were indeed vines, and there are even finer things to witness down here. If we squint, we might see those individual molecules and their atoms that make up the world, from here all the way back up. We can see how they are structured, how they are moving, potentially even how they're chemically reacting with each other and changing.

If we're lucky, here is where we might also see small carbon strands. When we take a closer look, we'll find that each strand is actually a hollow tube of rolled-up carbon atoms. These tubes can vary in size, but some may only be a dozen or so carbon atoms around. These are carbon nanotubes. We've had to shrink all the way down here to find them, because the smallest carbon nanotubes may only be a single nanometre across.

There are realms smaller than this; we still have a long way to shrink before we can see an atom's nucleus or zoom into the subatomic particles studied in particle accelerators such as CERN's Large Hadron Collider. But once we shrink beyond our current size, we'll become smaller than an atom, and our surroundings may become so strange that we won't recognise them as matter.

So, let's stop here. We are well and truly within the realm the physicist Richard Feynman once named 'the bottom'.

All the small things

A regular-sized human can't see the bottom, but this nano realm exists in parallel with the macro realm of everyday human life. The two worlds are inextricably intertwined. Without cellular mechanisms that operate on scales measured in nanometres, life as we know it on Earth simply would not exist. It is for reasons like this that acts of meddling with the nano realm can have enormous impacts back up in the world we can see.

Nanotechnology, then, is the craft of editing this parallel world of the very small.

This is also where our space elevator comes back into view. At first, it might be strange to fish for space elevator building material down in the nano realm. Common sense might tell us that a tube 100,000 times narrower than human hair should be hopelessly delicate. It might seem absurd to proclaim that something so ephemerally thin could ever be capable of withstanding forces that could cleanly pull a steel bar in two. On the contrary, a carbon nanotube's diminutive size is precisely what gives it incredible tensile strength.

A steel bar or a strand of spider silk is, really, a collection of many different molecules in aggregate. Look at

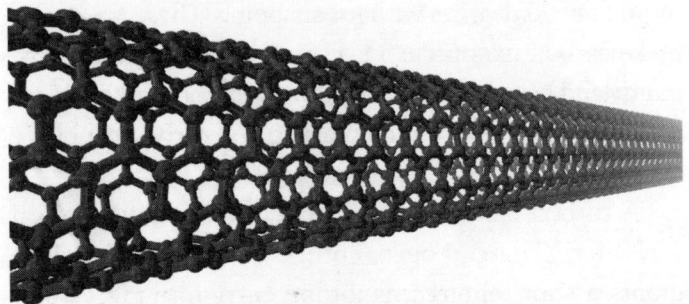

A carbon nanotube's structure.

either under a very powerful microscope, and you will see plenty of borderlines and defects and weak links where molecules meet. At the same time, a carbon nanotube's small width and simple structure mean that – at least in theory – it's straightforward to create nanotube molecules that span a room, unbroken from end to end.

The longest carbon nanotube we've ever produced measured around a staggering 50 centimetres. We shouldn't undersell this feat – that carbon nanotube is 100 million times as long as it is wide – but 50 centimetres won't take us into space. A carbon-nanotube-based space elevator is decades away, at least.

However, carbon nanotubes are useful for more than just building space elevators. Since a handful of researchers around the world discovered how to make them in the early 1990s, the carbon nanotube has grown into a starring role as one of the world's most studied materials.

It's easy to see how a carbon nanotube's strength might capture attention, but strength is only one of the carbon nanotube's curious properties. A carbon nanotube isn't just strong; it's also flexible, capable of bending more than 90 degrees without snapping. These properties make carbon nanotubes ideal for anything that must be sturdy and bendy at once. Fabric woven from carbon nanotube yarn is already on the market, and carbon nanotube aircraft parts may soon follow.

A carbon nanotube is flexible in more metaphorical ways, too. The carbon nanotube's ultra-narrow width grants a short enough nanotube entry into places that are otherwise too small to easily reach – inside human cells, for example. What's more, depending on the orientation of its atoms, a carbon nanotube can take either the properties of a metal (effortlessly conducting electricity a thousand times better than a copper wire) or a semiconductor like silicon (whose conductivity can be turned on and off, making it a potential material for futuristic electronics).

Carbon nanotubes are also amongst the world's best thermal conductors. You could stack a layer of intertwined nanotubes on a computer chip to wick away its heat, or you could mix a batch of nanotubes into water to create an enhanced cooling fluid. A carbon nanotube's tiny size also gives it more surface area per volume than virtually any other material. As chemical reactions rely on surface area, a chemist could speed up their reactions by utilising a carbon nanotube branch as the base for a highly effective catalyst.

Nanotechnology is a very new field, but it already has its emblems, and it's easy to understand why the carbon nanotube is one of them. So, what exactly is nanotechnology?

The word might conjure images of very tiny machines, crunching away in tiny corners where humans cannot see. A large part of nanotechnology does indeed deal with nano-sized machinery. We've been able to build basic machines ourselves; perhaps more impressively, we've already saved countless lives by adapting biology's nanomachinery. Many scientists and writers have gone far further, dreaming of far more sophisticated machines that can assemble any product molecule by molecule.

Yet there is more to nanotechnology than mere machines. Many nanoscientists, who carry out their work at this very tiny scale, might tell you that the nano realm's wonders are actually the materials that operate here. Carbon nanotubes are only one material on an ever-growing list: atom-thin sheets, specialised proteins lifted from cells, spheres that change colour as they shrink or grow. These materials are what really separate nanotechnology from the bulk world of human scales. Even if nanotechnology is still in its infancy, those materials already fill our world.

What, then, actually makes nanomaterials so unique? Why do carbon nanotubes work at all? While we answer those questions, what does a carbon nanotube look like when we unroll one into a flat sheet? When we do that, we'll find another material that has gained even more renown – graphene.

THE WORLD'S TINIEST ASSEMBLY GUIDE

2

In 2004, two Manchester University scientists named Andre Geim and Konstantin Novoselov were spending their Friday evenings sticking pieces of adhesive tape to chunks of graphite, then tearing them off again. If this seems like an activity that some chemists might devise to amuse themselves during a boring meeting, Geim and Novoselov were not merely trying to kill time. Instead, the two and their colleagues saw graphite and tape as a very serious means of scientific enquiry.

Six years later, Geim and Novoselov's acts of ripping away carbon flakes like skin onto a bandage would earn them a Nobel Prize in Physics.*

Geim and Novoselov were trying to make graphene – not to be confused with the graphite that was their starting point. If they could isolate graphene, they could

---

* They are two of many Nobel laureates we can thank for our knowledge of the nano realm. Others are coming.

isolate the thinnest material in the world. Other scientists had long predicted graphene's existence, and some had even caught glimpses of it, but none had ever been able to isolate it in its purest form. Geim and Novoselov believed they could do so with sticky tape. They also believed their tape could serve as a new way into the nano realm.

To understand what Geim and Novoselov were trying to create, we must first examine its fundamental building block: the carbon atom.

How to play with atoms

The idea of the atom was, of course, not new in 2004. A number of philosophers from across the ancient world suspected their universe was formed from invisible, indivisible particles. We can thank two such philosophers, the Ancient Greeks Leucippus and Democritus, for giving us the ancestor of the modern English word 'atom'.

Leucippus and Democritus knew nothing of modern chemistry, but even the chemist's picture of the atoms is centuries old. Around the turn of the nineteenth century, the chemist John Dalton proposed that the universe was formed from atoms, each one representing a chemical element. To this end, Dalton collated a primitive list of the elements, delineating a grand total of twenty, carbon included. When atoms join in different combinations, they form all the chemical compounds we see in the world around us. Dalton knew very little of the nano realm, and his picture was both incomplete (today's periodic table holds about six times as many elements as Dalton's list) and inaccurate (his list included 'elements' like 'lime' and 'soda' which we know today are compounds). Yet Dalton's general idea of the atom was sound.

Then again, the universe's indivisible building block is not actually indivisible. A century after Dalton, his successors like Ernest Rutherford and Niels Bohr worked out that atoms are made from even tinier parts, each millions of times smaller still: a positively charged nucleus (which we now know to contain positively charged protons and electrically neutral neutrons) circled by negatively charged electrons. Rutherford and Bohr also knew little of the nano realm; protons and neutrons lie in the realm of atomic and nuclear physics, largely beyond nanotechnology's reach.

By contrast, the electron is well within the nano realm's concern (if not in its own scale). A building block would be fairly useless without some way to actually link up with its fellow blocks. Electrons serve the role of atomic mortar. Within an atom, its electrons stack up within shells. Atoms are wont to fill their outer shells, and they'll bond to other atoms to make that happen.

Let's look at how this works in carbon. A neutral carbon atom has an outer shell with eight electron slots, four of which are pre-filled. Thus, the atom has four electrons to give other atoms and four slots to receive. Four electrons is a lot, allowing carbon to easily bond with hydrogen or nitrogen or oxygen or any number of other elements, forming the wondrous menagerie of molecules that make life on Earth tick. It also means carbon can easily bond with other carbon atoms in different ways.

Pure carbon comes in a number of forms, but in our macro world we are most familiar with two of them, and the two seem like the twin faces of a mercurial god. On the one hand, carbon can appear as soft and malleable graphite, which can shear away with the press of a pencil to paper. On the other hand, carbon can appear as durable diamond, the hardest material known to exist in nature, mounted on rings when not tipping industrial blades. So why are diamond and graphite so different?

The answer is in the shape of each material's innards. In diamond, all four of each carbon atom's outer electrons bind to its neighbours, locking the atoms on all sides in a cage – this gives diamond its legendary hardness. Graphite's carbon atoms only share three of their outer electrons, assembling in loosely stacked sheets. Each sheet's atoms are arranged in a honeycomb pattern, but each sheet is misaligned from the sheets above and below – when we drag a graphite pencil-tip across paper, sheets readily slide off.

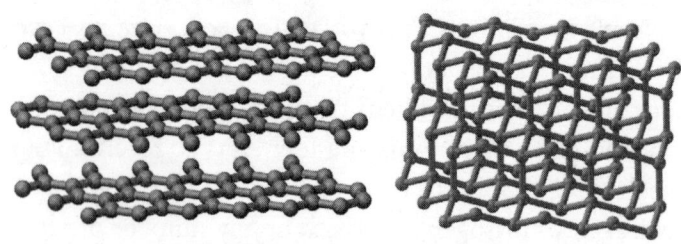

Graphite vs diamond.

Materials scientists call these geometries of atomic blocks and electronic mortar a material's crystal structure.* The world is resplendent with crystals. The atoms in table salt, sodium chloride, form tightly packed cubes. The atoms in solid ice are, like graphite, arranged in hexagons – this is why snowflakes form as six-pointed stars.

What, then, does a carbon nanotube's crystal structure look like? A carbon nanotube shares graphite's hexagonal form, but unlike a sheet in a graphite stack, a carbon nanotube's atoms are rolled up into a cylinder. The tube has no misaligned weak links. Hence, a carbon nanotube holds up to pulling with extraordinary strength.

The same structure gives the carbon nanotube a toughness that diamond lacks. A diamond's crystal structure may be a carbon cage, but the structure's 3D geometry means that diamond is easy to break apart. Diamond can withstand scratching better than any material, but swing a hammer at a diamond, and the stone will shatter.† A carbon nanotube's honeycomb has no such weakness.

---

* Not every solid is a crystal. Take amorphous solids, whose atoms are messily strewn about as if in a liquid. Glass is the most familiar amorphous solid; it forms as its molten form cools and hardens. Physicists don't actually fully understand what happens to glass as it cools. The lax and disordered structure of glass imposes upon the electron a very tortuous voyage. Glass is a very poor conductor and, instead, a very good insulator, which is why old telegraph wires relied on glass caps.

† Ironically, we might thank this imperfection for diamond's value. If diamond had a carbon nanotube's crystal structure, we wouldn't be able to cut one into an aesthetically pleasing jewel.

Shape is only part of the story. Unlike mortar, electrons aren't always fixed in place. Certain bonds can leave some electrons unmoored, free to flow through a material's crystal lattice. Flowing electrons make electricity; the more easily electrons can flow through a material, the better that material will conduct electricity.

We can see these two properties in diamond and graphite. Diamond's bonds keep its electrons all locked up, holding them from moving and making diamond an electrical insulator. On the other hand, while graphite's fourth electron isn't contributing to its strength, it's free to flow through its sheet, making graphite a wonderful conductor.

These are basic examples of how materials derive their properties from goings-on at the very tiniest scales. They are simplistic examples. Predicting what any material will look like is a formidable task that scientists are far from mastering, even armed with supercomputing clusters and machine learning.

These properties are also not unique to the nano realm. Everything in the universe that we can see, from an insect's exoskeleton to a skyscraper's steel facade, is made from atoms. However, not everything is made from the same number of atoms. The human body is made from some seven billion quintillion atoms. The far punier *E. coli* bacterium still contains eye-watering billions of atoms. What if we deal with objects made from, perhaps, a hundred atoms? What about a sheet of graphene like the one Geim and Novoselov wished to create, only one atom thick?

At those scales, nature starts to look a little different. We can thank a scientist from a time not long after Dalton for starting to show us how.

Strange science

Johannes Diderik van der Waals also knew little of the nano realm. Born in 1837 to a carpenter in the Dutch city of Leiden, van der Waals was an unlikely candidate for a career scientist. Instead, he left school aged fourteen to become a schoolteacher himself. In his spare time, he attended physics lectures at Leiden's famed university, but he could not enrol there, for his modest education had not taught him the Latin then required of all entrants. To his good fortune, the Dutch government did away with that ancient rule not long after. Van der Waals ultimately finished his doctorate at the age of 36, middle-aged even by that century's standards.

An unorthodox career path did not stop van der Waals from leaving behind an impressive legacy as a keen investigator of gases, a legacy that culminated in the 1910 Physics Nobel Prize. The idea that gases consisted of atoms and molecules was rather new in his day, but van der Waals modelled how such a gas would behave – every physics student now learns the maths he worked out. He reasoned, for instance, that passing atoms and molecules could push and pull each other.

Today, these pushes and pulls are known as van der Waals forces, and we know that liquids and solids are just as capable of experiencing them. With our modern understanding of subatomic particles, we know that van der Waals forces arise from negatively charged electrons sloshing around inside an atom, creating passing pools of electrical charge that attract and repel each other.

Compared to forces like gravity, van der Waals forces are pathetically weak at long distances, which is why we do not often find the floor repelling us off the ground. Measured on the scale of individual atoms, the picture is very different. For example, van der Waals forces allow atoms to stick together without forming formal bonds. In graphite clumps like the ones Geim and Novoselov had at their disposal, van der Waals forces are what hold the individual carbon sheets stacked together.

Van der Waals may have known little of the nano realm, but the nano realm is intimately familiar with van der Waals forces. For example, we don't usually see single carbon nanotubes in isolation. Instead, van der Waals forces entice nanotubes to join in multiples, bundled together like strands of yarn in a rope. This same adhesion lets us create 'multi-walled' carbon nanotubes, consisting of tubes stuffed within each other like the rings of a dart board. Van der Waals forces keep the nanotube layers fixed in place.

Chemistry works differently in the nano realm. A nanotube is a cylinder, and as a cylinder grows larger, its volume (which is dependent on the cube of the cylinder's radius) balloons much more quickly than its surface area (dependent on the square of its radius). Conversely, the smaller a cylinder, the higher its surface-area-to-volume ratio. This ratio is not merely a mathematical curiosity, because chemistry is a game of surface area. The greater the surface area, the more space that atoms and molecules can use to stick to – and react with – each other.

Poke around the nano realm some more, and we'll find oddities beyond van der Waals' comprehension. His work belongs to classical physics, which suggests that the universe operates according to elegant mathematical harmonies. By the time van der Waals passed away in 1923, he might have heard growing whispers about an alternate sort of physics. His colleagues had begun to realise that, at very tiny sizes, the idea of a symphonic classical universe breaks down under the hazy noise of quantum mechanics.

The heart of quantum mechanics, hence the field's name, is the idea that the universe is quantised: that properties like energy can't exist as just any number, but only as certain values. Imagine a staircase instead of a slope. This may seem like a minuscule difference, but when early-twentieth-century physicists dared to work out the maths, they discovered the distinction spawns all sorts of strange creatures who dwell deep in the murkiest corners of physics.

One of the best-known quantum creatures is wave-particle duality: that light, or another piece of the electromagnetic spectrum, can simultaneously take on the properties of a wave and of a particle. Another famed quantum beast is the uncertainty principle: the more certain you are of a particle's position, the less sure you are of its momentum, and vice versa. All manner of other spectres lurk in the bottom, announcing their presence when we venture down there. We'll meet others in due course.

Ordinarily, we are too large to see quantum beasts in our own world of bricks and skyscrapers.* On the other hand, if we zoom deeply into the heart of an atom, we'll enter a world dominated by quantum beasts. The nano realm occupies a middle ground between these two extremes. This is one of nanotechnology's special qualities: it can act as a sort of bridge between the strange world of these very small scales and our own mundane world.

Bridges like this have existed long before nanotechnology, long before Geim and Novoselov set out to create graphene, long before van der Waals, long before Dalton and long before any human ever imagined an atom at all.

Nature goes nano

The gecko lets us see the nano realm in action. A gecko can clamber up a tree trunk or a smooth building wall and make it look easy. It can relax on a ceiling, upside-down and apparently unbothered by gravity. It can even dangle from a glass pane by a single toe.

The gecko draws its abilities from interactions at this microscopic scale. Slide a gecko's foot under an electron

---

* It's not that the maths of quantum mechanics does not exist at human scales. Instead, the maths makes strange quantum phenomena so improbable that we can ignore them.

microscope, and you will see a forest of hairlike bristles – *setae*, in the biologist's parlance. One foot might sprout millions of setae; in turn, each seta branches into thousands of tinier nanohairs called *spatulae*. These bristles are made from keratin, the same material that forms human hair and nails, but the material alone is not enough (ask a human to dangle from a glass ceiling by their own hair).

Instead, it is the size and shape of the material that makes the difference. The tip of each spatula is about 100 nm across, narrow enough to summon van der Waals forces. These van der Waals forces can keep each tip tightly stuck to a surface. The adhesion of a single spatula's tip might be insignificant but remember that a gecko's foot is covered in millions and millions of spatulae. Combined, they give the gecko quite the sticking ability.

The animal kingdom hosts numerous other relics of the nano realm. Spider-feet also have van-der-Waals-enhanced hairs, allowing some spiders to support up to 170 times their own weight. A dragonfly's wings are pockmarked with nano-sized spikes that pierce any potentially infectious bacterium which tries to gain a foothold. The curious colourscapes of some butterfly-wings come from tiny particles embedded within, sending light rays scattering in strange directions.

We can find nanostructures in the plant kingdom, too. Take the lotus – not the car, but the flower that blooms atop a leafy pad floating upon lazy waters. From ancient devotees of the goddess Isis to the Buddhists of today, many around the world venerate the lotus as an emblem of purity. A raindrop that falls upon a lotus-leaf does not stick. Instead, it slides off and wicks away any dirt in its path. Even if a lotus-flower blooms in mud, its leaves will shimmer unblemished.

The divine purity of the lotus, much like the wall-clambering gecko or the eye-catching butterfly, may have a nanoscale explanation. Under an electron microscope, a lotus-leaf's surface will reveal a sprawling landscape of miniature bumps and hairs, glossed over with a wax coating. This landscape forms a natural water repellent. The wax keeps a raindrop spherical, prevents it from splattering; the bumps encourage it to roll away like a marble.

Nanostructures like these are – quite literally – merely the surface of the living nano realm. As we'll see in time, we'll find many more by taking a look deeper within. We'll find nano-sized natural machinery and nano-sized biological factories.

Geckos are cute and lotus leaves are certainly aesthetically pleasing, but we can do more than just admire their nano-parts through a microscope. When scientists and nano-architects start building materials of their own, they often look to nature for inspiration. For example, some materials scientists have created glass that mimics the lotus's miniature rugged terrain and sends raindrops rolling away. Soon, we might find lotus-inspired glass in windows or windscreens that clean themselves.

Meanwhile, plenty of researchers have devoted their time to making completely new nanostructures.

The thinnest materials in the world

The craft of building nanostructures is perhaps the largest part of nanotechnology today. We've constructed all kinds of nanostructures in all sorts of shapes and proportions. Any structure will do in the nano realm, so long as at least one of its three dimensions can be measured in nanometres.

The most basic sort of nanostructure is the nanoparticle. If something can be measured in nanometres on all three dimensions, it's a nanoparticle. Satisfying that basic rule, a nanoparticle can be and may be made from any substance. We use metallic nanoparticles (iron and gold are common), organic nanoparticles (made from carbon-based materials like lipids and cellulose) and nanoparticles in earth tones (made from materials like clay).

Nanoparticles might not seem like the most glamorous things in the world, until you can see what their size lets them do. They can alter a substance's basic properties, strengthen a material or change its colours.

We think of gold, for instance, as a spectacular regal yellow, but shred the precious metal down into tiny nanoparticles, and a shard of gold can turn red or black in the light. The same is true for other types of nanoparticles. Take titanium dioxide, a substance found in everything from paints to food colourings. Although large particles of titanium dioxide are white, nanoparticles can be transparent and absorb ultraviolet light instead; you'll often find titanium dioxide nanoparticles in sunblock. Other nanoparticles can absorb light and 'upconvert' it into a different colour. Chinese scientists recently used such nanoparticles to make contact lenses that allow humans to see in the infrared.

Because nanoparticles are so small, they can go into places that are otherwise nigh impossible to enter – human cells, for example. We've also discussed another advantage of a nanoparticle's size: the smaller it becomes, the more surface area it has per volume. As we've mentioned, chemistry is a game of surface area; the more surface area you've got, the more room you have to host chemical reactions. Hence, nanoparticles make excellent catalysts. Nanoparticles of sparkling platinum, for instance, help with the hard work of producing hydrogen in fuel cells and breaking down pollutants in catalytic converters.

Some specialty nanoparticles have special names. Hollow out a nanoparticle, and you've a nanoshell that you can fill. Stretch a nanoparticle out into an elongated cylinder, perhaps three to five times as long as it is wide, and you have a nanorod. Wrap a nanorod into a circle, and you have a nanoring. We've made nanocubes, nanocones, even nanobowls. Nest multiple nanoshells within each other like Russian dolls, and you get multi-layered 'nano-onions'. Engineers have grown minuscule spindling nanoflowers and constructed nano-helices that can spindle through a fluid with the motion of a very small propeller.

More than just exercises in nano-geometry, each of these nano-shapes has its intended purposes. Nanoshells and nanobowls are excellent packaging for other substances – if nanoparticles can enter human cells, then a drug-containing nanoshell can deliver its payload right into a cell's interior. Nanoflowers could indeed sprout in a little nano-meadow, but the expansive surface area of a nanoflower's petals makes it excellent at absorbing pollutants from the environment or other substances. Researchers at the University of Queensland recently built nickel-based nanoflowers that tested blood for signs of pregnancy complications.

Those, then, are the sorts of structures that fit into tiny three-dimensional boxes, but nanostructures also come in more spaghettified forms. Stretch a nanorod further, stretch it into a strand thousands of times thinner than a human hair, and you've a nanowire. Make the nanowire hollow, and you've a nanotube. These long lines can get very lopsided indeed – we've mentioned that the longest-ever carbon nanotube is 100 million times as long as it is wide. Carbon is the most famous nanotube, of course, but we've also made nanotubes from other materials, like silicon, boron nitride and gallium nitride.

Then there are nanostructures that look like very flat sheets. At its very thinnest, a nanostructure can be a layer as thin as a single atom. This is one of chemistry's ultimate size limits: it's as thin as most materials can possibly go. This is also the sort of material that Andre Geim and Konstantin Novoselov wished to create: graphene, a two-dimensional sheet of carbon.

Finding graphene

So, we return to Manchester in 2004.

Geim and Novoselov were not the first scientists to think about making 2D carbon. Scientists had long known that graphite was made from many atom-thin layers of carbon, bound by van der Waals forces. It's easy enough to separate the graphite sheets, but when you do so – as you do when you drag a pencil across paper – you almost always slough away multiple sheets at once. What if you isolated a single sheet instead? What would that sheet look like?

As early as the 1940s, some scientists had predicted that a single layer would act rather differently from 3D graphite, but as we've said, predicting a material's properties isn't easy. Scientists needed to get their hands on the real thing. Six decades later, no one in Manchester or anywhere else knew whether they would ever be able to touch the real thing. Many scientists believed an atom-thin layer would be too unstable to do anything more than crumble at the slightest touch.

Hoping to resolve this dilemma, Geim gave a doctoral student named Da Jiang a lump of graphite and a task of creating the thinnest sheet possible. Da's first attempts didn't resolve anything. He tried to grind down the graphite but could not manage to get a graphite shard anywhere near the ultimate thinness. (As it happened, Geim had accidentally given Da a type of graphite that was far too dense.)

Another of Geim's students, overhearing this dilemma, pointed out that the operators of certain microscopes known as scanning tunnelling microscopes, or STMs, often used graphite to adjust their equipment. The microscopists stuck adhesive tape onto a graphite chunk, then tore the tape away, leaving behind a smooth surface useful for calibration. The discarded tape, then, carried some ultra-thin carbon detritus. Geim scoured through laboratory bins for all the graphite-stained tape he could find. The discarded graphite still wasn't an atom thick, but it was much thinner than any other carbon Geim's lab had managed to intentionally create.

This was when Geim recruited Novoselov with the intent of replicating the tape method. They first found that sticking tape to graphite and ripping it away once might take away 100 or 10 layers. Impressively thin, but Geim and Novoselov were still short (or long, as it were) of their goal. They then realised that, if they repeated the process several times, they could slowly whittle down

Graphene's structure.

the layers until they had only one. After several tries, they had their single – they had graphene.

Graphene's name is very similar to graphite, but graphene could hardly be less like its soft, three-dimensional cousin. On the contrary, graphene is the strongest material known to humankind. Without vertical weak links to burden it, all of graphene's bonds are of the strong carbon-to-carbon variety. These bonds give graphene a carbon-nanotube-like tensile strength when you pull at it, yet graphene is extremely flexible.

What's more, like the carbon nanotube, the carbon atoms in a graphene sheet still have their unbidden fourth electrons that are free to roam. Graphene thus conducts electricity about 100 times better than copper. So, too, are both carbon nanotubes and graphene sheets excellent heat conductors. In essence, heat is a measure of how atoms vibrate, and how solids transmit heat is dependent on the strength of their bonds. Because carbon nanotubes and graphene both have such sturdy bonds, both materials transmit heat ten times more effectively than graphite and a hundred times more than iron.

Nevertheless, graphene and carbon nanotubes are not identical. The very act of unrolling a nanotube into a sheet, even if we don't do anything else to the atoms, makes an incredible difference on the material's properties. Place two paper sheets on a table: one flat and another rolled into a tube. Logically, the flat sheet will have far more contact with the surface below. The same is true for graphene. Every carbon atom in a graphene sheet is open on two sides. A batch of graphene can have anywhere from two to 25 times more surface area than the same mass of nanotubes. If you want to make ultrathin carbon stick to another material, graphene does it better.

Shape gives rise to other differences as well. Carbon nanotubes and graphene may share a crystal structure, but they do not necessarily conduct electricity in quite the same way. There are multiple ways of rolling carbon atoms into a tube. This leads to the difference we briefly discussed in chapter 1; different configurations conduct electricity in different ways. A pure graphene sheet, on the other hand, only comes in one configuration, and that configuration is one of the best electrical conductors in the world. (As we'll see in chapter 5, this distinction has very real relevance for engineers trying to make carbon-based electronics.)

There are other differences, too, which we'll describe as we slowly examine what graphene can do.

Building from the bottom up

It's well and good to talk about atom-thin sheets and nanotubes, but how do you create them? How do you construct other nanoparticles, other nanomaterials or anything in the nano realm at all?

In creating graphene with tape and manual labour, Geim and Novoselov show us a common method of nano-construction. Their method may seem crude, but there is a logic behind it. Geim and Novoselov began with a large chunk of raw material – bulk material, in the nanotechnologist's parlance – and broke it down, steadily shaving away bits, until they wound up with the nanostructures they wanted.

Researchers who work in the nano realm call this sort of thing *top-down* manufacturing. It's a perfectly reasonable method, especially if the nanoparticles you want are easy to create by whittling them off. But the nature of nanotechnology opens the door for an alternative approach. We could dispense with bulk material entirely and start from the bottom. We could assemble atoms or molecules, one at a time, into the nanostructure we want, assuming we find a chemical process that will actually make the material we want (which is certainly not a given).

Nanotechnologists call this alternate approach *bottom-up* manufacturing. Bottom-up methods have gradually supplanted top-down methods for the highest-end graphene, because bottom-up methods can produce graphene sheets with greater size and far fewer impurities than top-down manufacturing can manage. One typical method is to grow graphene with a technique called chemical vapour deposition (CVD). This method starts with a base, known as a *substrate*, which shares graphene's honeycomb structure. (Copper and silicon dioxide are common substrate choices.) Then, under the proper heat and chemical conditions, carbon atoms build up one at a time as a layer upon the substrate, stuck on by van der Waals forces. CVD is a far cry from the Manchester group's tape; it often requires high temperatures and specialised equipment, which can make it rather costly.

Carbon nanotubes can be made from the bottom up, too. We usually can't roll graphene sheets up into nanotubes,* but we can make nanotubes using a similar CVD process. We can also take a lump of graphite, vaporise it into carbon atoms and let those atoms settle back into a structure as they cool off. When they do, they often sprout as budding nanotubes. Scientists call this sort of formation *self-assembly*.

The ability to build from the bottom up, very literally, is another one of the features that makes nanotechnology so special. We are, in essence, using the basic building blocks of the universe – atoms and molecules – as building material. It might seem like stacking bricks and mortaring them into a house, but there is a slight size difference between placing another brick in a wall and placing another atom in a crystal structure. Imagine constructing a brick by assembling its billions of atoms one at a time.

We've talked about nanostructures in nature, and Earth's biology can similarly show us examples of bottom-up manufacturing in action. Cells construct proteins from the bottom up by meticulously piecing them together from smaller molecules, like assembling parts into an aeroplane. If that sounds much like the machinery of a factory, it is machinery that is vital to supporting life on Earth. Indeed, some nanotechnologists believe

---

* We can, however, do the reverse. Years after Geim and Novoselov's initial efforts, some researchers succeeded in making graphene by 'slicing' a carbon nanotube down its axis and unrolling it into a flat sheet.

we may soon be able to harness machines like these for ourselves.

Nanotechnology at its most sophisticated is one of science's younger fields. Under that name, it has only existed since the last decades of the twentieth century. However, nanotechnology has been with us for centuries longer.

RACING TO THE BOTTOM 3

Lost in the upper levels of the British Museum, surrounded by Byzantine jewellery and treasures from Sutton Hoo, you'll find a singular glass cup. Lit from the outside, the relief of the legendary Greek king Lycurgus of Thrace, which covers the cup's body, is coloured jade-green. But place a lamp inside this so-called Lycurgus Cup, and its colour will switch to an eerie blood-red, as if it belongs in the house of a vampire.

There is probably no vampire's curse on the Lycurgus Cup. Instead, the Cup bears tiny particles of gold and silver, some as small as 50 nm in diameter, suspended within its glass. When light reflects off the Cup's surface, these precious metal nanoparticles stay hidden. When light passes through the glass, the nanoparticles announce their presence by scattering away the shorter-wavelength blue light and leaving the longer-wavelength red light for human eyes to see.

The Lycurgus Cup is fashioned from what a modern materials scientist would call dichroic glass, which changes colour depending on how you light it. Such dichroic glass is readily available today, used in both art and electronics. However, dichroic glass was somewhat less readily available 1,700 years ago in fourth-century Rome. Historians don't know if the ancient artisans who made the Cup knew they were giving it colour-changing properties. Whatever the case, the Lycurgus Cup is an example of how humans were using nanotechnology millennia before the carbon nanotube.

When visiting the British Museum, you can find more examples of the technology at work just a few steps away in the mediaeval Islamic rooms. Several centuries after the Western Roman Empire fell, potters in Iraq began gilding their wares with a magnificent metallic lustre. Their secret ingredients were mixtures of vinegar with copper- or silver-containing powders. If a potter applied metallic slurry to their glazed ceramic before firing, their piece would emerge from the kiln glistening as if plated with gold. Over the following centuries, this technique – known as lustreware – spread from Iran to Al-Andalus. Some potters even thought of lustre as a kind of alchemy. They were not entirely wrong; wittingly or not, their process turned the copper and silver into reflective nanoparticles.

Meanwhile, over in early mediaeval Europe, artisans discovered they could colour glass by adding certain powders to their molten glass in the furnace. Adding silver nitrate and gold chloride turned glass vivid yellow and bejewelled ruby red, respectively. These particular stains, we now know, arose from nanoparticles; silver nitrate and gold chloride formed colour-changing nanoparticles in the glass. Some of Europe's great cathedrals can thank nanoparticles for their rainbowed windows.

Tricks of the light? Perhaps, but another example shows us that pre-industrial peoples tapped nanotech not just to transform the colours of their materials, but also to strengthen them – the fabled Damascus steel. For over a millennium, metalworkers in that ancient city forged blades with striking zebra-patterned bands. Damascus steel blades gained a legendary reputation for sharpness and durability; tales were told across Eurasia of otherworldly swords that could split stones in two without dulling.

The recipe for Damascus steel was lost before the Industrial Revolution, but wash a Damascus steel sword with hydrochloric acid and place it under an electron microscope, and it will reveal some of its secrets: carbon nanotubes and iron-carbon nanowires. These nanostructures give Damascus steel its distinctive bands. Modern materials science tells us that these same nanostructures can harden and strengthen steel.

Skilled as the smiths of Damascus or the glassmakers who blew the Lycurgus Cup must have been, they almost certainly knew little of molecular bonds or carbon nanotubes. Nanotechnology as we know it would have to wait until well into the twentieth century.

A physicist's dream

Many nanotechnologists today will tell you that their field began one evening in late December 1959, when the American physicist Richard Feynman took a stage in Southern California.

Scientists today may remember Feynman as one of the twentieth century's great particle physicists; others might know him for his efforts to broadcast science to the world. On this night, Feynman had stood before a gathering of his fellow physicists to talk about neither of those things. Just over two years after *Sputnik*'s launch, many in Feynman's audience no doubt had daydreams of the realms that existed beyond Earth's atmosphere, celestial realms now tantalisingly within reach. Feynman had not come to talk about space, either. Instead, Feynman began to speak of a very different realm, in a very different direction.

'What I want to talk about is the problem of manipulating and controlling things on a small scale,' he said. He wondered if it could be possible to write the Encyclopaedia Britannica in a typeface so small that the whole text could fit on the head of a pin. He wondered how the pieces of an infinitesimally small automobile might fit together. He wondered if engineers could make computers with 'sub-microscopic' parts. He wondered if patients might one day swallow a 'mechanical surgeon' that could operate inside the tight confines of a single blood vessel. He wondered if we could create machines at this scale, transforming atoms and molecules into billions of nano-sized lathes.

Feynman spoke of these small scales as if they were ripe with boundless opportunities. 'There is plenty of room at the bottom,' he said, giving the talk its best-remembered title.

So, did nanotechnology as we know it today emerge from Feynman's brain that evening? In reality, probably not. Feynman did not mention how anyone was actually meant to reach the bottom; as Feynman himself acknowledged, the technology to enter the nano realm on such grand scales simply did not exist in 1959.* The early history of nano-exploration was far more sluggish and scattershot, driven by researchers at varied points around the globe, many unaware of Feynman's interest and few sharing his grand visions. It is often the nature of scientific progress to be meandering and ponderous, its goal uncertain before it actually arrives.

Nonetheless, progress was happening. Some of the earliest forays into the nano realm took the form of polymers: materials made from small molecules arranged into long chains. Humans had long used naturally occurring polymers like rubber, but by mid-century, chemists had found ways of swirling molecular components together

---

* Actually, some of it did exist. That night in 1959, Feynman offered two prizes of $1,000 each (about £8,500 in 2025), one for the first person to print a book-page on a 1/25,000 scale and another for the first person who could make an electric motor fit into a 1/64-inch cube. That second prize was claimed only a year later by an electrical engineer named William McLellan using conventional electronics that already existed, much to Feynman's bemusement. The first prize did have to wait until 1985.

such that they assembled into synthetic polymers from the bottom up – a key nanotech technique, as we've discussed.

Elsewhere, by the 1960s, scientists had developed tools that used beams of light or molecules to deposit atoms in films as thin as a few nanometres, or to etch shapes whose sizes could be measured in nanometres. These nano tools may seem like arcana, but materials scientists still use them today; they likely helped create the silicon chips that power your computer or mobile phone.

Indeed, tiny electronics seemed to be the first of Feynman's prophecies to come true. Many researchers in the 1960s and 1970s began turning their minds to the nano realm precisely because they could already see electronics shrinking with their own eyes. Year by year, the components on circuit boards were becoming smaller and smaller, drifting further and further away from human eyesight. It was only a matter of time before electrical engineers would need to reckon with the murky nano realm and its oddities.

These issues were on the mind of a Tokyo University of Science professor named Norio Taniguchi when, in 1974, he published a review paper describing the techniques that engineers could use to chisel away individual molecules or atoms from ultra-precise electronic components – techniques he called 'Nano-technology'. Taniguchi envisioned 'nano-technology' as ways to finish parts: a far cry from nanometre-print encyclopaedias. Yet just as nanotechnologists today may credit Feynman for creating their field, they sometimes credit Taniguchi for coining the word.

Even then, nanotechnology (which soon lost its hyphen) did not immediately catch the attention of the scientific world. Scientists needed to make several additional advancements first.

Touching atoms

You cannot enter the nano realm with the sort of microscopes you will find in a school biology classroom. Not even the best-equipped lab in the world can venture far with a visible light-based microscope. Nature throws up a physical limit: the wavelengths of light itself, ranging from 700 nm (for red) to 400 nm (for violet). Try to see smaller objects, and interference increasingly muddles the light. Under normal circumstances, the sharpest visible-light microscope can't discern objects smaller than around 200 nm in size: a blurry line known as visible light's diffraction limit.* So, you'll need to find another vehicle.

You might try the far tinier electrons. By the second half of the twentieth century, the electron microscope had come into vogue. Electron microscopes come in a few different designs, but all of them fundamentally work by shooting a high-energy electron beam into a sample. By the 1970s, the best electron microscopes had resolutions

---

* It's actually possible to see past this limit, but doing so requires optical trickery that didn't exist yet in the 1970s.

down to around a nanometre, a size that successfully puts nanoparticles back in the playbill.

It's hard to understate how useful electron microscopes have been, including for scientists trying to study nanomaterials, but electron microscopes are still not quite precise enough to spot individual atoms. Furthermore, they have less in common with biology-classroom tools than with ray guns: an electron beam can easily vaporise a sample.*

Fortunately for microscope samples everywhere, microscope-makers around this time had devised a gentler alternative. Rather than firing an electron beam, you could pierce the nano realm's veil with a very, very sharp tip. Bring this tip close enough to an atom – about a nanometre, say – and electrify it with a slight voltage. This can entice an electron to jump from the atom to the probe.

An electron ordinarily shouldn't have the energy to make such a jump, but recall that shadowy quantum beasts stalk the nano realm, and it is here that one comes out to play. This beast is called quantum tunnelling, and it allows particles to make seemingly impossible moves. Imagine kicking a football into a brick wall. Classical physics and common sense alike tell us that the ball should inevitably bounce back. The maths of quantum mechanics instead tells us that the football is not a single point but a wave-like cloud of many possible points. If the wall is

---

* Several years ago, an imaging physicist told me that using an electron microscope is something like trying to look at light through a beer bottle.

thin enough, there's a chance that the cloud's edge might cross it, giving the ball a chance to pass right through. (In practice, with an object the size of the football, this isn't likely to happen once in the lifespan of our universe.)

In the same mathematical way, an electron near a metal probe can make the seemingly impossible jump across the gap to the probe's point. Move the point over a surface like a raster scanner, count how many electrons you draw away, and you have an effective way to map atoms. A machine that works like this is called a scanning tunnelling microscope, or STM.* Two IBM engineers, Gerd Binnig and Heinrich Rohrer, created the first STM in a Zurich laboratory around 1980.

An STM is a wonderful machine, and even in the early 1980s, the scientific community understood how useful it could be. An STM is also a very delicate machine. It requires a near-perfect vacuum, for any matter at all will disrupt the jumping electrons. It also requires that any heat be kept to an absolute minimum, which in the 1980s meant cryogenic deep-freezing to near absolute zero. Furthermore, while an STM's tip must be very sharp in order to reach the resolution of atoms, it cannot be too sharp, else it will snap off at the slightest touch.

Even as Binnig and Rohrer travelled to Stockholm to share the 1986 Physics Nobel, it was all but inevitable that

---

* Years later, the STM would play an unwitting part in the discovery of graphene. STM operators used graphite to calibrate their equipment, preparing it by sticking tape and tearing it away. Geim started searching through bins for the detritus from graphite.

their invention would give its users splitting headaches. Such was the experience of fellow IBM engineer Donald M. Eigler. Working in California, Eigler could scarcely have been further away from Binnig and Rohrer's Zurich base. Eigler wanted to use an STM for taking atom-sized images, but like a batch of energetic children, his atoms simply refused to sit still. Whenever Eigler brought the STM's tip close to an atom, the whole atom tended to jump out of plane and cling to the probe.

Attempt after frustrating attempt passed, until Eigler noticed a pattern: his atoms only seemed to misbehave when he brought his probe very close to an atom. When the distance shrunk to a few atomic diameters, the atoms slipped out of place. Knowing this, Eigler could learn, little by little, how to control their movement. His annoyance faded as he realised that an STM could do a bit more than just take images. He could use the STM to pick atoms up, move them around and pin them elsewhere.

In November 1989, Eigler and a colleague named Erhard K. Schweizer chilled a nickel canvas to nearly absolute zero and began to ornament it with xenon atoms. The pair picked up and placed down 35 atoms, one by painstaking one, until they had recreated three 5-nm-tall letters: spelling out the name of their employer.

Publicity stunt? Nano-advertising? Most likely. Still, Eigler and Schweizer had just created the world's first work of nanoscale art. For the first time in recorded history, scientists could rearrange individual atoms like marbles in a sandbox. If somebody wished to fulfil Feynman's wish of writing an entire encyclopaedia on the

The IBM miniature.

head of a pin, they now theoretically had the writing system to do it. Playing with individual atoms was not yet easy – Eigler and Schweizer needed an entire day to write just three letters – but it was now possible.

The STM became the nano-architect's first real method of manipulating the very, very bottom. It was not the only discovery to herald that the nano realm was open for exploration. Another had come from space.

Digging for buckyballs

When the future Sir Harry Kroto was a young man, he was one of the millions who visited Montréal for Expo 1967. World's fairs like it tend to double as showcases for experimental architecture – past fairs had given London the Crystal Palace and Paris the Eiffel Tower – and Montréal's was no exception. Towering over the Expo was a 60-metre-tall geodesic dome, a spherical shell formed from a mosaic of interlocking polygons. The dome, designed by the architect and noted geodesic dome enthusiast Buckminster Fuller, housed the expo's American pavilion.* To visitors like Kroto, the dome must have seemed like a vision of an interplanetary colony. It was a vision that stayed in Kroto's mind.

Nearly twenty years afterwards, Kroto was a scientist in Sussex contemplating an interstellar puzzle. Kroto was a chemist by trade but had also become a part-time skywatcher as of late. His puzzle did not lie in the stars themselves so much as in the vast expanses between them. In recent years, radio telescopes had picked up the fingerprints of curious serpentine carbon molecules speckled across interstellar space. No one was certain where those molecules had come from.

Kroto suspected that these carbon chains might form in red giants. Most stars gather their energy by fusing hydrogen atoms, smashing them together to create helium and a considerable amount of energy, but red giants are aging stars that have run out of useable hydrogen fuel and resorted to smashing helium instead. Fusing helium atoms tends to create carbon, and Kroto thought it logical that the immense heat of a celestial atmosphere could fashion those carbon atoms into chains. If Kroto wished to test this hypothesis, he and his colleagues would need to recreate a red giant on Earth.

Fortunately, Kroto knew of a suitable device. Across the Atlantic, at Rice University in Texas, Kroto's fellow

---

* Unlike many world's fair pavilions, and despite a 1976 fire that ravaged its interior, Fuller's dome is still around. Today, the dome houses the Biosphère, an environmental museum.

chemists Richard Smalley and Robert Curl held the keys to a particularly powerful laser. Kroto proposed to aim that laser at a lump of graphite, heating its carbon atoms up to thousands of degrees and pulverising them into carbon vapour. As the vapour cooled back down, its atoms would clump together and rain out as new carbon molecules, unbidden from graphite's crystal structure. Watching scientists could collect the molecules and take a census of their masses. If their count included masses matching those of Kroto's carbon snakes, Kroto would have his evidence.

It was a bemusing and odd idea for an apparatus intended for studying how materials formed, not for simulating stars, and an entire year passed before Smalley and Curl gave in to Kroto's request. Nevertheless, when Kroto, Smalley and colleagues finally fired up the laser in the autumn of 1985, they found Kroto's carbon chains within days. This story may well have ended there, but – as so often happens in science – in the process of solving one puzzle, another emerged.

Beyond the 6-to-8-atom carbon snakes, the laser had seemingly spawned scores of a completely unknown molecule that weighed in at 60 carbon atoms. These mystery molecules were far heavier than anything the chemists had expected to count. No one knew where else it could be found. No one even knew what a 60-carbon-atom molecule looked like. The group had one clue in the fact that this mystery molecule did not seem to react with its surroundings, hinting that its carbon atoms formed some sort of closed shape – perhaps a sphere.

At the mention of a sphere, Kroto envisioned himself back in Montréal, back in the shadow of Buckminster Fuller's geodesic dome. Armed with paper cut-outs and tape, the chemists set about arranging model atoms into Fuller's domes. Their result was a patchwork of pentagons and hexagons resembling the black-and-white pattern of an old-fashioned football. This new molecule gained the architect's name: buckminsterfullerene. Or, as it is perhaps better known: the buckyball.

For centuries, scientists had taken for granted that carbon came in two Earthly forms: soft and supple graphite

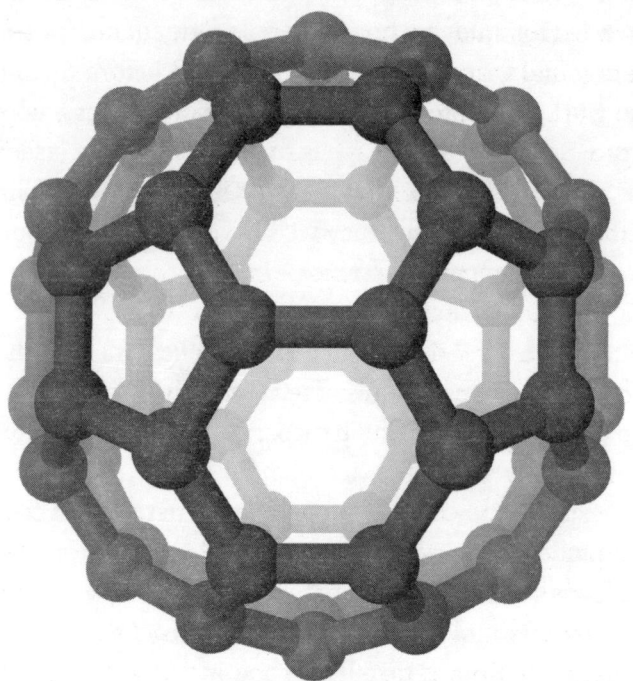

The buckyball.

against hard and durable diamond. The buckyball threatened to entirely upend that picture. Many chemists were sceptical of its discovery, but by 1990, another group had determined how to make piles of buckyballs and thoroughly test them, and there could be no doubt that the buckyball was real. Kroto, Smalley and Curl eventually won the 1996 Chemistry Nobel for their trouble.

The buckyball became many chemists' introduction to the nano realm, and scientists of all stripes came from all corners in search of more buckyballs and fellow travellers (known collectively as fullerenes). The original Rice experiment had also found traces of a 70-atom variant. Many of the other groups trying to explore fullerenes after them found variants ranging from 32 to 80 atoms.

One of those followers was Sumio Iijima, an engineer working for Japan's NEC Corporation. Iijima had the opportunity to experiment with an arc discharge: a device not unlike the one in Texas, if you swapped out the high-powered laser for intense jolts of electricity. Like the buckyball's discoverers, Iijima used his machine to vaporise carbon atoms, but when Iijima's carbon cooled back down, he noticed something that the buckyball-purveyors had not: thorny crowns formed from little carbon needles.

When Iijima looked more closely with an electron microscope, he noticed that the needles were actually hollow carbon tubes, a few nanometres across. Iijima wasn't the first scientist to observe carbon nanotubes, but after he published his results in *Nature* in 1991, he became credited as the first to knowingly create them.

By the early 1990s, human scientists knew well how to work in the nano realm. How much more could humans build down there? For some, the potential seemed all but limitless.

Nanotech goes big

Kim Eric Drexler had spent virtually his entire career dreaming of realms beyond his own sight. The first of those realms was outer space. When Drexler was a student at MIT, he worked on imagining the sorts of infrastructure we'd need in order to permanently settle in Earth orbit.* Drexler's 1977 master's thesis was effectively a blueprint for a solar sail,† a device that would push a spacecraft by catching the Sun's solar wind. Key to Drexler's design was an ultra-thin coat of aluminium, one that could reflect light without weighing down the sail. Drexler posited a coat less than 100 nm thick.

Then, not entirely unlike Kroto, Drexler's attention gradually fell from the heavens to realms far below. His dreams turned to nano-creations more elaborate than aluminium films. Somewhere along the way, Drexler

---

* Drexler worked with Gerard O'Neill, a physicist today best known for proposing that space-dwelling humans might live in miles-long cylindrical habitats.

† This book isn't really about space exploration, but we can look at solar sails to get an idea of science's timescales. A solar sail was first tested in interplanetary space when the Japanese probe IKAROS hoisted one on a mission to Venus in 2010.

had dusted off the script of Feynman's decades-old talk. What separated Drexler from Feynman was that, by the early 1980s, scientists had actually started working at the bottom level that Feynman had envisioned. Drexler began to wonder just how much scientists could do down there.

In a 1986 book entitled *Engines of Creation: The Coming Era of Nanotechnology*, Drexler imagined a nano realm filled with machinery. Drexler pointed out that, in nature, biological proteins already performed many machine-like functions. Earth's cells used protein machinery to snip and edit DNA. Other proteins attached to bacteria and served as arms or propellers. Still others could replicate and assemble themselves.

But Drexler's dreams went far beyond mere biological nanomachines. He imagined that, with the hands of human scientists, proteins would eventually give way to 'second-generation nanomachines' that could assemble atoms into 'almost any reasonable arrangement'.

We've talked about bottom-up assembly, but bottom-up assembly then and now relies on lucking into chemical methods which only work with certain materials. The nanomachines that Drexler envisioned would be able to pick and move atoms of any kind like pieces of a construction toy set. Drexler's book imagined that these 'molecular assemblers' would fundamentally reshape how humans interacted with their world. With these nanomachines, Drexler promised, we could build anything from houses to rocket engines from scratch, from the bottom up, one molecule at a time.

Engines of Creation landed on the public's bookshelves at the right time, stirring up dreams of the sort of future that nanotechnology could bring. The sort of nanomachines and nano-sized factories that Drexler imagined were still very far away, but by the last decades of the twentieth century, other sorts of nanotechnology were fast accumulating. This was the era when nanotechnology, as scientists understand it today, began to trickle out of the lab. Looking back, we might consider the 1990s as the era in which nanotechnology made a grand entrance into the wider world.

Textile manufacturers began embedding nanomaterials between fabric fibres to make clothing resistant to stains. Sunscreens had long relied on compounds like titanium dioxide and zinc oxide to separate human skin from ultraviolet radiation, but these compounds tended to mark the skin with ungainly white streaks; now, sunscreen-makers began reshaping those compounds into streak-free nanoparticles. Within just a few years, basic nanomaterials found their way into paints, cleaning sprays, deodorants, tennis racquets and computer displays. Nanomaterials have filled human materials for centuries, but now the average consumer could buy products they knew contained nanomaterials.

Other nanomaterials found more clinical uses. In 1995, American regulators approved an ovarian cancer medication named Doxil. At its core, Doxil wasn't a new drug; its active ingredient was doxorubicin, a chemotherapy drug that had been around for decades. Doxorubicin causes a great deal of collateral damage in the form of soreness, hair loss, nausea, peeling skin and other symptoms familiar to anyone who has ever known a cancer patient. Doxil's real advance, then, was to repackage doxorubicin within organic nanoparticles that could seamlessly enter individual cells. Doxil is often considered the first nano-enhanced drug to gain the favour of government regulators. It was not the last. By decade's end, doctors in both the US and EU could prescribe nanomedicines for diseases ranging from hepatitis to multiple sclerosis.

Other scientific advancements showed off what the newly accessible nano realm had on offer. Computer chips had so drastically increased in density that their components started to plunge into the nano realm. Physicists had begun producing tiny dots that tugged at quantum reins to change colour with their size. Chemists had begun to actually turn molecules themselves into machine parts. Around that time, somebody noticed that the newly discovered carbon nanotube might possess the immense strength required to build an elevator cable to geostationary orbit.

Major governments established national nanotechnology institutes and began pouring eye-watering billions into funding them. In the business world, major corporations followed suit, and hundreds of startups emerged with promises of technology that would reshape human society in a new Industrial Revolution. Many of their claims echoed Drexler's ideas of nanotechnology-sized factories. For a time, 'nano' became technology's hottest buzzword.

As for Drexler himself, he became nanotechnology's public figurehead, engaging in public debates over whether his molecular assemblers could really be built. As his dreams of a nanotech-based future caught an audience, so too did his warnings of a potential nanomachine apocalypse. Essay-writers and other critics who watched these debates began to publicly fret that nanotechnology was more energy behind an impending technological tsunami that would sweep humans away.*

All this was the picture of the world that Andre Geim and Konstantin Novoselov might have seen painted for them when, in 2004, they announced they had isolated graphene for the first time. Their newly discovered wonder material seemed to be yet more proof that an all-encompassing nanotechnology revolution was just around the corner.

As of some two decades later, nanotechnology has not made humans obsolete.

The course of scientific progress is often meandering and unspectacular. Molecular machines – and space elevators – were never going to materialise wholesale in just a few years' time. The venture capitalists who funded

---

* The zenith of this sentiment may have come from a scientist named Bill Joy. In April 2000, Joy penned an article in *Wired* magazine entitled 'Why the Future Doesn't Need Us'. Within it, Joy prophesised that nanotechnology – when coupled with robotics and genetic engineering – threatened to make humans an endangered species. Joy considered these technologies even more dangerous than atomic weaponry and the like, because while nukes were the purview of large governments, nanotech and robotics were being pushed by large companies.

those hundreds of startups perhaps realised this, and they turned their eyes elsewhere. Look at the popular science press of *today*, and you might see nanotechnology depicted as a technological revolution that failed.

That depiction isn't really accurate, either. We may not have Drexler's molecular assemblers, but again, nano-technology is just as much a science of materials as it is a science of machinery. Those materials have not stopped improving. Just as the nano realm surrounds us, often unnoticed, these advancements are quietly reshaping our world.

In fact, where Drexler imagined that the biological machinery of proteins would serve as humanity's first introduction to the world of nanomachinery, many of nanotechnology's greatest advances have played out in the medical world – including one that has surely saved millions of lives.

A FANTASTIC VOYAGE 4

Imagine a scenario: a patient has entered a coma. There is a blood clot in the patient's brain, one buried in a corner that well-established medicine cannot reach. Fortunately, there is a new experimental technology: a machine that can enter the brain and despatch the blood clot with a well-placed laser. The complication is that the human body is very good at identifying things that do not belong. This machine must reach its destination and carry out its mission before the guard lymphocytes of the patient's immune system can identify it and destroy it.

Such was the plot of the 1966 film *Fantastic Voyage*, a dream of nanotechnology that was released decades before nanotechnology went mainstream. Admittedly, it's not quite nanotechnology as we would recognise it. (For one, the machine has a crew, who must shrink down and enter the body like submariners. We do not yet have the technology to shrink humans down to the size of a blood vessel. For another, their mission plays out rather like a mission to outer space, complete with mission control watching them from an elaborate version of a hospital operating room.)

What's true is that the sizes involved in this mission are well within reach of nanotechnology today. We see the craft traverse the patient's tiniest capillaries, which we know can be as small as 10,000 nm across. We see the craft sail past red blood cells, each one about 6,000 nm across. And we see the patient's immune system start to mark the craft for termination by coating it with antibodies, which are much smaller: as small as 10 nm. These are sizes small enough for nanoparticles – or, indeed, carbon nanotubes.

What could we do with abilities like these? One of the first things we may wish to do is to take a look around.

Darkening in the glow

It is no simple task to see inside a live human. The body is filled with secrets, concealed corners and inaccessible hiding spots, and, for most of human history, the body's interior may have been as mysterious as the Moon. Physicians had few reliable ways of seeing inside the body without cutting open the skin. The knowledge that anatomists did collect over the centuries, from ancient Egyptian priests to mediaeval Islamic doctors to the likes of Leonardo da Vinci, largely came from dissecting human cadavers. Alas, dissection is not particularly effective (or comfortable) for the living.

Hence, the history of medical imaging is usually said to have begun in 1895, when Wilhelm Röntgen discovered the X-ray.* Röntgen was playing with an electron beam when he noticed that, even if he blacked out the tube with thick paper, objects in his lab lit up in bright fluorescence. Röntgen figured he was working with an unknown type of ray. In one of his follow-up experiments, Röntgen cast these newly dubbed X-rays through his wife's hand onto a photographic plate behind. When he had the plate developed, his wife's bones appeared before his eyes. Röntgen had found a way to peel away his wife's skin.

Bones in X-ray images are actually silhouettes. They appear because bones are better at absorbing passing X-rays than the soft tissue around them, which they can do because bones are denser than most of the body. Even in Röntgen's day, physicians racing to get their hands on primitive radiography machines understood that if they wished to see a human body's squishy organs, they would need to fill them with a densifier, a 'contrast agent'. If you've ever had a barium meal examination, you've ingested a contrast agent. The chalky drink you put down contains barium sulphate, a compound that's also denser than the body's surroundings. The barium sulphate coats the walls of the GI tract as it passes through, and coated surfaces cast an X-ray shadow.

---

* We don't actually know how this happened, because Röntgen pointedly had his lab notes destroyed.

Nanoparticles, as it happens, make excellent contrast agents. They are small enough to sneak into individual cells, and they are endlessly customisable. By tuning their size, we can control how they behave: larger nanoparticles might be more visible, but the kidneys can filter smaller nanoparticles out of the body more quickly. Nanoparticles can be made from different materials and wrapped in different biological markers to direct them on different courses through the body.

When it comes to X-ray contrast agents, few nanoparticles are more desirable than those made from gold. Outside the body, jewellery collectors covet gold for its lustre; inside the body and under X-rays, gold is, by contrast, valuable for its lack of a shine. Like barium sulphate sludge, gold nanoparticles absorb X-rays more effectively than their surroundings, leaving behind distinct shadows on resultant images. Gold is easy to work into all kinds of shapes and easy to outfit with guiding molecules.

One such guide is a molecule called RGD peptide. It's ubiquitous in the body; it acts as a cellular adhesive, attracting cells and helping them stick together. By modifying the molecule, we can attract different sorts of cells. We can, for example, modify it to make it more recognisable to cancerous cells. We can coat a gold nanoparticle with this modified RGD peptide and turn it into a tumour-seeker. If there is an occluded tumour somewhere hidden in the body, such nanoparticles flock to it and settle inside. Their opportunity comes when a radiologist activates an X-ray beam and starts to scan the body. Assuming everything has gone according to plan, the gold nanoparticles turn tumours into little shadows in the image.

X-rays were our first way of seeing into the body, but they were not the last, and their alternatives, too, have nanoparticles to assist them. By the middle of the twentieth century, doctors began pinging bodies with ultrasound – frequencies of sound too high-pitched for the human ear. To make certain body parts more sensitive under ultrasound, scientists can deploy nano-sized bubbles filled with gas. When sound waves wash over the nanobubbles, they vibrate and twinkle like stars in the resulting ultrasound image.

The world's most modern hospitals host MRI machines, which pierce the body's veil by shrouding it with a powerful magnetic field. This field pulls the natural magnetic fields generated by the protons in the body's water and fat molecules all into alignment, then lets them go. Protons in different pieces of the body take different times to settle back to normal, and these differences show up in the MRI image, allowing physicians to differentiate them. Nanoparticles can shorten this time, essentially brightening a certain tissue in the image. MRI-enhancing nanoparticles are often made of iron oxide or from gadolinium, one of the rare earth materials that inhabit the forgotten basement of the periodic table.

So, engineers can certainly tune nanoparticles to seek out particular medical conditions in the body. The obvious question, then, is whether we can use that ability to treat those same conditions. As it happens, this provides another application for the carbon nanotube.

Treating the untreatable

Glioblastoma is one of cancer's most fearsome forms. For many patients, it may as well be a curse of death. Traditional cancer treatments are not particularly effective for this brain tumour. Even if a glioblastoma tumour is removed by a surgeon's scalpel or glassed by chemotherapy or radiation, the cancer tends to return. On average, a glioblastoma patient survives just around a year after diagnosis. Fewer than 10 per cent of patients survive longer than five years.

What physicians could do, before long, is send in a batch of properly outfitted carbon nanotubes. One by one, the nanotubes would settle into their proper places inside the brain tumours. These nanotubes are stuffed with a magnetic iron oxide filling. When researchers turn on a magnetic field, they catch the nanotubes in their grasp. As the researchers manipulate the field, they can incite the nanoscalpels to swirl around like food processor blades, destroying the brain tumours from within.

Such is a treatment that University of Toronto researchers, in 2023, successfully tested inside cancer-stricken mouse brains. These nanoscalpels are a long way from reaching the hospital, but less spectacular nano-sized cancer treatments for humans have slowly trickled through clinical trials and approval since the 1990s.

We mentioned one such treatment, Doxil, in the previous chapter. Doxil is a chemotherapy drug filling a liposome: a shell made from lipids, a group of organic molecules that includes fats and waxes. The human immune system is very good at identifying things that do not belong, but lipids *do* fit into the body, and the right liposome can ensure that the medical cargo is properly protected. By modifying the structure to include different types of lipids in different proportions, biologists could tune their liposomes to go one way or another. One lipid mix might be ideal for reaching cells in the spleen; another might be ideal for floating to cells in the lung.

Carbon nanotube blades and liposomes are by no means the limit of potential nano-medications in the clinic today. Other approved nano-drugs hitch rides into the body on orbs of albumin, a protein found in blood plasma. Still others enclose their drugs into nanocases of iron oxide, which allow them to be guided from the outside like those carbon nanotube blades. In the laboratory, researchers are busy finding ways to pack medicines into golden nanorods, nanostars or nanoshells.

These applications are mostly aimed at cancer, but physicians can use the same nano-toolbox to target other diseases. Nanoparticles could latch onto HIV, which is often exacerbated by its viruses being prone to hiding in reservoirs where they are difficult to reach. Scientists in Singapore have designed gadolinium-based nanoparticles that can travel through the bloodstream and land in hidden masses of arterial plaque, which the nanoparticles can illuminate in an MRI. Then, since plaques are more acidic than the rest of the blood, each nanoparticle contains an outer shell specifically designed to erode in the presence of the acidity and release a drug inside.

Nanoparticles have another advantage against diseases like glioblastoma, because they're able to pierce one of the body's best-fortified thresholds: the barrier between the bloodstream and the brain. As arteries rise into the brain and fan out into a forest of capillaries, each tuft of blood vessels is wrapped by this blood-brain barrier. Under normal circumstances, this barrier helps keep the brain free of toxins – but it also keeps out beneficial medicines. Package a drug in a nanoparticle coated with the proper molecules, on the other hand, and its small size and camouflage allow it to pass unimpeded. This ability could allow nanoparticles to treat neural diseases like Alzheimer's and Parkinson's, on top of brain cancer.

No matter the disease, these sorts of treatments will become more sophisticated. Their means of getting nanoparticles to their destinations are already chemical and biological and, in time, may become mechanical as well. MIT researchers have tested mechanical pills, each one about the size of a fingernail, which a patient can swallow; after a pill descends down the oesophagus and descends into the digestive system, it latches onto a bit of stomach lining or intestine and injects a drug straight through a microscopic needle. Other labs have developed an optical fibre smaller than the width of a human hair that can fit into some of the body's smallest blood vessels. Today, gadgets like these are too large for the nano realm, but engineers expect that to change.

The Toronto lab's carbon nanotube nanoscalpels also passed right through the blood-brain barrier. Moreover, they passed the blood-brain barrier under the control of magnet-wielding researchers from the outside. By one definition, that feature turns the nanoscalpels into primitive nanobots. As we'll see in a later chapter, they are far from the only sort of nanobot that researchers have created.

Then again, some engineers of the nano realm would argue, why use custom nanomachines when the cells of the body already contain certain nanomachines of their own? A budding part of nanotechnology is the art of turning cells for our advantage.

Cell couriers

In chapter 1, we briefly discussed the idea that cells are akin to nano-factories, containing nano-sized assembly lines. We might not think of cells as factories, but without their ability to mass produce proteins, life on Earth would not exist. Virtually every biological process on the planet, from reproduction to digestion, relies on proteins. Cells – or, more precisely, certain organelles inside those cells – are responsible for churning out those proteins.

The average adult human is fashioned from some 30 trillion cells. Not all produce proteins; about five in six are red blood cells, which lack organelle-machinery and act as goods wagons that transport oxygen. Even so, that leaves every human with several trillion tiny biological factories. It's possible to think of the human body as a vast industrial landscape.

For those cells which do produce proteins, the chief machine of their assembly line is the ribosome, a biological fabricator that's generally 20 to 30 nm across. The ribosome takes basic parts – small molecules known as amino acids, of which there are twenty varieties – and welds them together into proteins. How, then, does this fabricator know which amino acids to assemble into which shape? The cell contains instructions written in genetic code, in the form of DNA, 2 nm across.

There is a middle step between DNA and ribosome. A cell stores DNA in its nucleus, in a sort of central archive, separate from the ribosome; even if the ribosome could touch it, the nucleus contains a codex of the entire genome, far too much information for the fabricator to use at once. Instead, a cell must copy the relevant section within the genome and send it to the fabricator. A cell's courier of choice is another molecule: messenger RNA (mRNA). Think of mRNA as a snippet of industrial automation code, a little tangle of nucleic acids that tells a cell which proteins to actually produce.

It's a process that also offers us an intriguing opportunity to interfere. If we can customise mRNA of our own, then send it to particular cells, we could repurpose their assembly lines to make proteins we want. We could, for example, programme the human body to produce pieces of a virus. The outer shell of a virus is studded with distinctive spike proteins. With well-placed mRNA, we could command human ribosomes to create such spike proteins.

This sounds like an act of biological subterfuge, but it's a useful act of biological subterfuge. As we've seen, the human body is extraordinarily good at recognising things that do not belong inside it, and viral spike proteins certainly do not belong. Soon enough, the body's security systems activate and despatch lymphocytes to neutralise the threat. The immune system thus learns how to do the same when a genuine spike protein enters the body. What we've got is a vaccine, but unlike a traditional vaccine, it contains no part of an actual virus.

Reprogramming is far easier said than done. Biologists have known about mRNA since the 1950s, but they couldn't use it to their advantage until decades later. When they first began injecting mRNA in clinical trials in the 1990s, they usually watched helplessly as their mRNA degraded before ever reaching a cell. Again, the human body is extraordinarily good at recognising things that do not belong. The very same human immune system is wont to attack the mRNA code snippet before it even reaches the cell. Even after the mRNA does enter the cell, it must pass the cell's endosomes, another set of machines that act like postal sorters and filter out unsolicited junk mail.

So, proper packaging is key to the whole operation. The obvious choice, considering the scales at hand, is a nanoparticle: a liposome. Unlike Doxil and its fellow medications, these liposomes don't contain drugs, but information. We can package the mRNA inside a liposome. As a bonus, van der Waals forces keep the mRNA molecule bound to the liposome's inner shell, bracing it and providing additional protection during its rough ride through the body's elements.

Liposome-packers had reason to believe their packaging could work. Their colleagues had successfully used liposomes to wrap a smaller but related molecule known as small interfering RNA (siRNA). If mRNA is a set of instructions, siRNA is an emergency stop – bringing a protein-fabricator to a halt. siRNA is a good way to make cells cease producing harmful proteins, a common result of genetic disorders. In 2018 and 2019, American regulators gave their approval to two drugs that used liposome-wrapped siRNA to treat two rare diseases caused by faulty protein product.*

Acquiring approval for a vaccine is no small feat. A vaccine must prove its quality through a series of clinical trials, usually at least three in number, each potentially lasting several years. Rightfully, scientists and regulators want to monitor trial subjects for quite some time to ensure that a vaccine actually proves its quality as a protector. By the 2010s, some mRNA vaccines had entered clinical trials for diseases like rabies and chikungunya (a neglected tropical disease spread by mosquitoes), but it seemed like these were only the first steps for a technology that might pay off in a generation. For the most part, mRNA vaccines and their liposome nanoparticles remained in the pharmaceutical backdrop: an interesting idea, but not one that seemed positioned to make any significant impact.

Normal circumstances disappeared very quickly after the World Health Organisation declared Covid-19 a global

---

* Hereditary ATTR amyloidosis and acute hepatic porphyria, to be precise.

pandemic in early 2020. In the vaccine development frenzy that followed, the liposome-riding mRNA vaccine was one of the many vaccine concepts that got a metaphorical shot in the arm. Thanks to the thankless work that researchers had done in prior years, it was also one of the most successful.

Within less than a year, hundreds of millions of mRNA vaccine doses had begun to roll out across the world. It is possible that this act of nano-hacking is the most influential application of nanotechnology to date.

Genes on the cutting room floor

In 1987, Osaka University medical researchers combing through the genetic code of the infamous bacterium *E. coli* found a strange bit of DNA. The organism's genome seemed to contain a strange recurring pattern. Imagine reading a book only to find your text repeatedly interrupted by an identical out-of-place paragraph.

Now, imagine turning to a different book in the same genre and finding the same paragraph, the same strange recurrences. In the following years, similar patterns revealed themselves in the genes of countless other single-celled organisms. Scientists named the cryptic interrupts 'clustered regularly interspaced short palindromic repeats', or CRISPRs for short.

The first major clue to the system's purpose came when scientists realised that the genetic code between a bacterium's CRISPRs didn't come from the bacterium at all. The CRISPRs were library shelf markers for a genetic archive of particularly fearsome viruses known as bacteriophages. The word 'bacteriophage' is derived from the Ancient Greek for 'bacteria-devourer', but a bacteriophage doesn't consume bacteria so much as hijack them. Shaped like a nano-sized war machine, a bacteriophage latches onto a hapless host and injects it with a payload of malicious DNA. This insidious gene contains instructions for spawning more bacteriophages. If it infiltrates a bacterium's genome, it corrupts the host's machinery into a bacteriophage assembly line.

Faced with a menace worthy of classic *Doctor Who,* what is a bacterium to do? Some bacteria have evolved an ingenious defence method: recognise the malicious DNA and excise it. The exact mechanism varies from creature to creature; in some bacteria, it works like this. If a bacterium survives a bacteriophage attack, the bacterium saves the invasive DNA in its CRISPR memory. When the next attack comes, the bacterium can 'remember' its attacker and deploy a DNA-snipping protein to slice the threat into pieces.

This was what biologists like Emmanuelle Charpentier knew by 2011, when Charpentier met a biochemist named Jennifer Doudna in a Puerto Rican café. Charpentier's group had studied *S. pyogenes,* the bacterium that inflicts humans with scarlet fever; they knew that *S. pyogenes* snipped its DNA with a protein named Cas9, a hand-shaped molecular cutter about 5 nm long and 10 nm across. What they didn't know was how exactly Cas9 knew which piece of DNA to cut.

Charpentier and her colleagues suspected that Cas9 relied on RNA strands they had spotted, RNA strands that seemed to match the CRISPR code. This was why Charpentier had come to Puerto Rico seeking the aid of Doudna, a renowned RNA specialist. Doudna agreed. Over the following months, they and their colleagues were able to puzzle together the pieces. When a bacterium recognises a bacteriophage attack, it copies the relevant genetic code from the CRISPR DNA archive into a corresponding RNA strand. The bacterium then equips Cas9 with this RNA strand, which guides Cas9 to the matching malicious DNA.

What made this experiment truly remarkable was what Charpentier and Doudna's group did next. The researchers tried outfitting Cas9 with different 'guide RNA' snippets of their own choosing, each corresponding to different points in the bacterium's genome. Each time, Cas9 snipped the bacterium's DNA where the researchers desired. The researchers had repurposed a bacterium's own mechanisms into a pair of customisable 'genetic scissors'.

Genetic scissors make genetic editing quite elementary. Suppose you have a gene you want to deactivate. All you must do is equip the 'genetic scissors' with a matching snippet of guide RNA, something that is fairly routine to code using the modern biologist's toolset. The Cas9 will snip the DNA accordingly, and the cell will try to repair the cut itself. Ideally, this deactivates the cut gene. Biologists soon realised they could upgrade their system by supplementing the scissors with a genetic template. Not only would Cas9 snip the DNA, but the cell would also use the template to manufacture new DNA to edit into the cut.

Charpentier, Doudna and their peers did not invent the art of genetic engineering; biology labs have birthed genetically modified organisms since the 1970s. CRISPR is different due to its precision. Old established genetic engineering techniques largely meant injecting the new gene into a cell and allowing the cell to randomly insert it somewhere within the cell's genome. CRISPR, on the other hand, lets genetic engineers control precisely where they want to edit.

We knew about other DNA-cutting tools before CRISPR: other proteins with fleshy names like zinc-finger nucleases (ZFNs) and transcription activator-like effector nucleases (TALENs). But CRISPR is simpler and cheaper; guide RNA is easy to produce and easy to use, even for biology labs of modest means. CRISPR makes editing a genome nearly as elementary as editing a film.

Although biologists found their genetic scissors in an infectious bacterium, CRISPR gene editing will work with virtually any organism. In 2020, a Japanese biotech company began selling tomatoes enriched with the dietary supplement GABA – the first CRISPR-enhanced plants to enter the market. The following year, Japan approved the sale of pufferfish and red sea bream that had been fattened up with CRISPR's aid.

These approvals are merely the tip of a far larger iceberg of CRISPR-enhanced plants and animals: albino watermelons; tomatoes coloured yellow, pink and purple; catfish with alligator genes that boost their disease resistance; livestock with loosened controls on muscle growth; mosquitoes stripped of the genes that make them liable to carry malaria-bearing parasites. Again, edits like these may have been possible for decades, but CRISPR has sped up a traditionally slow process. Moreover, CRISPR easily allows biologists to edit the genes of a living organism – even a human.

In November 2023, British regulators approved a new method for treating sickle-cell anaemia and beta thalassaemia: conditions in which the body does not properly produce haemoglobin, the protein that red blood cells use to carry oxygen. Before CRISPR, treating these conditions typically meant relying on regular transfusions of blood from a healthy donor for life. After CRISPR, doctors can take a patient's stem cells, reactivate a gene that instructs cells to make haemoglobin, then reinject the modified cells back. If all goes well, the repaired cells proliferate through the body's bone marrow, where blood cells are manufactured.

Faulty red blood cells might be the first medical condition to get approved CRISPR treatment, but if scientists have their way, many more could soon follow. As of this writing, trials are under way applying CRISPR to everything from HIV to type 1 diabetes to lupus to, indeed, cancer. Charpentier and Doudna watched all of this happen, having won the 2020 Nobel Prize in Chemistry for their work.

Just three years later, another band of researchers would win the same prize: not for a machine from the cellular world, but for creations that harness beasts from the even more shadowy quantum world.

ELECTRONS AT THE NANO GATES 5

The selection of a Nobel laureate is, by tradition, a watertight process. An entire year before a prize is awarded, that prize's committee invites nominations from a select group of notables – for the science Nobels, that is an exclusive club of past laureates, members of the Royal Swedish Academy of Sciences, professors at Nordic universities and several other honoured scientists from around the world.

Their list of nominees is kept secret by statute of the Royal Swedish Academy. If you wish to know whether you've been nominated for a Nobel prize in a certain year, you have two options: wait half a century, for the statute of secrecy expires after 50 years; or clear the committee's year-long judgment for the Academy to contact you with the good news, shortly before the official announcement that you're a laureate.

Only then can a name be released to the public, and only the victors' names are revealed. The Academy takes its charge so seriously that, in the pre-digital era, it greeted reporters by despatching couriers with sealed envelopes. Only after the official announcement, only after its courier received the Academy's signal by telephone, could an envelope's contents be unveiled.

The Internet's ever-creeping reach makes this level of secrecy increasingly difficult to keep up. On the chill autumn day in 2023 when that year's Chemistry laureates were to be announced, several Swedish media outlets received a press release announcing that the Academy had awarded the prize to three scientists of the nano realm: Aleksey Yekimov, Louis E. Brus and Moungi G. Bawendi. The release was premature; the winners had not yet been notified, and they weren't to be announced for another several hours.

The slip made headlines around the world, but few scientists were especially surprised by the announcement itself. Yekimov, Brus and Bawendi had won their award for unearthing and crafting one of the nano realm's brightest jewels: the quantum dot.

Quantum dots

To understand the quantum dot, we must first understand the quantum beast that lurks within.

Quantum mechanics tells us that, just as waves of light can behave as particles called photons, so too can matter – anything from electrons to cricket balls – behave as waves. This also means that electrons and cricket balls each have wavelengths. Physicists call these *de Broglie wavelengths*, after Louis de Broglie, the French-aristocrat-turned-physicist who worked out the maths.

Like most quantum animals, de Broglie wavelengths tend to be meaningless in the human-scale world. The de Broglie wavelength of the cricket ball from a fast bowler's arm, for example, is about a trillionth of a trillionth of a trillionth of a metre – not especially relevant to the waiting batsman!

But de Broglie wavelengths are not always a mere mathematical curiosity. An object's de Broglie wavelength depends on its momentum: in other words, its mass and its speed. The lighter an object and the more slowly it moves, the longer its wavelength. Replace the cricket ball with an electron, which has very near a trillionth of a trillionth of a trillionth of a cricket ball's mass. An electron moving at slow enough speeds might have a wavelength closer to billionths of a metre, nanometres. In the nano realm, de Broglie wavelengths do matter.

This quantum beast is actually a rather well-known one.* As early as the 1930s, in the days when quantum mechanics was newly discovered science, physicists predicted that we might see oddities begin to emerge when we made small enough nanoparticles. But those physicists knew far less about the nano realm than we

---

* The quantum dot is actually an example of the so-called 'particle in a box' problem. It's one of the first problems that physics students encounter when they're introduced to quantum mechanics, not least because it is actually solvable in a reasonable amount of time.

do today, and they'd have to wait before actually seeing those oddities.

Four decades later, in the city then called Leningrad, a physicist named Aleksey Yekimov was poring over coloured glass. Scientists like Yekimov knew that doping glass, infusing it with another compound, filled the glass with nanoparticles of that compound which tinted the glass different colours depending on how the glassmaker treated it. To Yekimov, this didn't make much physical sense – why should the same chemical create different colours?

Seeking clarity, Yekimov made his own glass and doped it with copper chloride. When he examined the copper chloride nanoparticles, he realised that the glass's colour closely tracked the size of a nanoparticle. The smaller the nanoparticle, the bluer the light it absorbed.

This is because a nanoparticle essentially acts as a cage for its electrons. The de Broglie wavelengths its electrons can take, translating into the wavelengths of light they can absorb, are limited by the size of their cage. Smaller particles give their electrons less room to breathe, constraining them into higher energies, forcing them to absorb shorter-wavelength blues and violets. Inversely, larger particles absorb lower-energy, longer-wavelength yellows and reds.

Later, it became clear that the process worked just as well in reverse, that quantum dots can emit light at the same wavelengths. You'll often see this side of quantum dots demonstrated with a rainbow of jars, each containing a glowing liquid in a different colour. Each jar contains a quantum dot colloid; dots of a different size suspended in a liquid. The smaller the dots, the closer their emitted light to the blue end of the rainbow.

Simply by changing a particle's size, you can elicit a different response from the quantum beast caged within. This is how physicists might expect atoms or subatomic particles to behave, but a quantum dot is instead made from many atoms. When you see a glowing quantum dot jar, you're seeing quantum mechanics with your very eyes. It's an example of how nanotechnology can bridge two worlds that are otherwise entirely separate.*

The Iron Curtain split the scientific world just as it did the political one, and though Yekimov published his research in a Soviet journal in 1981, word of his discovery didn't immediately reach the West. In the New Jersey suburbs, chemist Louis Brus knew nothing of Yekimov's glass when, in 1983, he noted the same colour-follows-size effects in nano-sized crystals of cadmium sulphide suspended in a solution. Brus wanted to control the size of these nanoparticles, but he had no easy way to do so.

---

* You may recall that the gold nanoparticles we've encountered, such as in the Lycurgus Cup, could also change colour to red or black. Quantum dots aren't to blame, but the cause is a different quantum creature known as surface plasmon resonance. If light strikes a nanoparticle at the right wavelength, an electron on the particle's surface can absorb the light, thus changing the particle's colour. The exact wavelengths of light that get absorbed vary with the nanoparticle's size, shape and surrounding materials. Controlling surface plasmon resonance is trickier than controlling quantum dots, though many are trying.

In 1988, a young researcher named Moungi Bawendi encountered this dilemma as a postdoc in Brus' lab. Five years later, it was Bawendi, now at MIT, who solved it. Bawendi's technique relied on injecting cadmium selenide into a hot solvent. The cadmium selenide formed crystals, but the injection cooled the solvent, stopping the crystals from growing. However, Bawendi realised that if he heated the solvent again, the crystals would continue to grow. They would grow for as long as the solvent remained heated. By adjusting this time, researchers could precisely control the nanoparticles' size – and their colour. Bawendi's method is still used to manufacture quantum dots today.

This sequence of discoveries earned Yekimov, Brus and Bawendi the 2023 Nobel Prize in Chemistry. By then, quantum dots had begun to exit the lab.

The best-known use of quantum dots is probably the QLED display. Each green and red pixel in a QLED screen is fashioned from quantum dots sized for its respective colour. Turn on the screen, and you activate a blue backlight, rendering blue pixels in the empty cells and colouring the quantum-dot-filled cells green or red. We're not quite good enough at making useful blue quantum dots, the smallest of the lot, but if they improve, displays of the future could cut out the backlight entirely and instead light up their quantum dots with electric current.

Quantum dot cameras are now on the market. As light enters a quantum dot camera, it passes through a layer of quantum dots and jolts electrons if its wavelength is right. The most sophisticated filter incoming light through multiple layers of quantum dots, each in a different size and sensitive to different colours. Quantum dots are more sensitive than the sensors that digital cameras traditionally use, and the image quality is much better too. Engineers are interested in using quantum dots for night vision goggles and automated cameras more sensitive than anything self-driving cars use today.*

Scientists are excited about what quantum dots can do for them in the lab. Biologists who once relied on dyes to colour certain cells, like tumours in lab mice, can instead insert fluorescent quantum dots to light them up in higher resolution. Chemists can sprinkle quantum dots into a beaker and let the quantum dots' electrons stir up reactions in the surrounding liquid. Some engineers have proposed lining structures like bridges or airframes with a coat of quantum dot paint. When an inspector runs an ultraviolet laser across the paint, the quantum dots ought to glow green. If they shine redder or bluer instead, it's a sign that the quantum dots have changed size – that the structure beneath has been stretched or squeezed.

We'll encounter some more uses of the quantum dot soon enough. In the meantime, the quantum dot can illuminate a path that may lead us through the bottomlands to computer parts that are as thin as an atom.

---

* Today, it's relatively easy to make quantum dot cameras that are especially good at detecting near infrared, which makes these cameras ideal for seeing at night and seeing through adverse weather.

The saga of semiconducting silicon

You can fashion a nanoparticle from virtually anything, as we've seen, but not every nanoparticle will become a quantum dot and interact with light as quantum dots do. The key to creating a quantum dot is to ensure that its electrons have the right energy.

This is a matter of physics. The outer electrons of a material can slot into certain energy levels, or bands. Electrons tend to rest in a lower band, or valence band; electrons in this lower band stay in an atom's own orbitals. Give such an electron enough energy and it will jump into the upper band, or conduction band; electrons in a conduction band are free from an atom's grasp and can float around.

The amount of energy needed to transition an electron is called the band gap, and it varies from material to material. The breadth of a material's band gap tells us about how that material conducts electricity. Wide band gaps are the hallmark of an electrical insulator; elevating an insulator's electrons across its gaping gap takes impractical amounts of energy, so an insulator's electrons tend to stay fixed in place. At the other extreme, metals and other electrical conductors have overlapping bands, leaving no band gap at all; a conductor's electrons tend to flow freely.

Quantum dots lead us to the materials in between – semiconductors. These materials have band gaps, but their gaps are narrow and easy to span.

Quantum dots must be made from a semiconductor. When light strikes a quantum dot, it pushes one of the quantum dot's electrons into the conduction band; when the electron falls back down to the valence band, it emits light. LEDs operate by a very similar means – as the electrons in an LED fall from the upper band to the lower band, they emit their lost energy as photons of light. The wider an LED's band gap, the more energy zips away, and the bluer its light.

LEDs and quantum dots are only a very tiny slice of the semiconductor world. Indeed, the quantum dot Nobel laureates were already well aware of what else semiconductors could do.

Not only did the three carry out their work in two countries divided by the Cold War, but their places of work also mirror their respective sides. Yekimov studied glass at the S.I. Vavilov State Optical Institute, a Soviet state research facility founded under Lenin and by Yekimov's time renamed for a physicist whose brother died in a Stalin-era prison. On the far side of the Iron Curtain, Brus and Bawendi conducted much of their work at Bell Labs – the research arm of the corporate syndicate that controlled America's telephone network for much of the twentieth century.

The Cold War coincided with Bell Labs' golden age, when researchers there introduced the world to everything from solar cells to lasers to the Big Bang's microwave afterglow.* Near the zenith of that golden age,

---

* Yes, the first people to measure the cosmic microwave background were two Bell Labs physicists, Arno Penzias and Robert Wilson. The two were working with a radio telescope on the Bell Labs campus. When they eliminated all the interference from the telescope that they could manage, they detected a strange all-encompassing hiss. At first, Penzias and Wilson thought they might have detected pigeon droppings piled up in the antenna's horn-shaped receiver. They had actually found something far more ancient than pigeons.

about three decades before the first quantum dots blinked on, Bell Labs researchers were the ones who mastered the best-known semiconductor: silicon. In the 1950s, Bell Labs birthed the first silicon transistors.

A transistor is a piece of circuitry that lets you control an electrical current, or switch it on and off, by changing the electrical voltage applied to special electrodes. You can do this in a few different ways, but the one that caught on – also known as the FET – relies on an electrode separated from the current's channel by a thin insulating layer. (The insulation is important; without it, current would leak, rendering your transistor useless.) If voltage is applied to this electrode, known as the 'gate', it generates a modest electric field that 'opens' the channel and lets current flow. Switch off the voltage, and the gate will close.

Silicon is especially useful as a semiconductor, because it can easily be doped, or modified with atoms of another element. Doping silicon allows you to pick at the element's band gap. You can let electrons have energies slightly above the lower band (known as p-type doping) or energies slightly below the upper band (known as n-type doping). When you place the two types next to each other, you create an electronic one-way road that only allows current to flow in a direction you choose.

You can chain pairs of n-type and p-type transistors together to build a logic circuit. In itself, this feature does not make the transistor unique. You could hypothetically build logic circuits from the electrical switches in your house.* The earliest electronic computers relied on vacuum tubes, pill-shaped glass bulbs that were in computer engineering vogue before the transistor came along. But a transistor's voltage can be changed in a fraction of the time you need to reach over and switch off your lights, and the transistor is far more energy-efficient than the vacuum tube or any other alternative.

The silicon transistor in particular had an even greater advantage: silicon was easy to mass produce, shrink down, stick on a chip, and mass produce some more. It's estimated that *sextillions* of transistors have been manufactured since the 1950s. Humans have produced more silicon transistors than any other object in all of recorded history.

Over the second half of the twentieth century, engineers pushed the silicon transistor along a one-way journey towards the bottom. The first silicon transistors were easily visible to the naked eye.† A bulky 1950s calculator fitted several thousand transistors into an apparatus the size of a desk. Just twenty years later, the Intel 4004 – often considered the first microprocessor to reach the market – fit about the same number of transistors onto a chip the size of a postage stamp. Twenty years later still, Intel's Pentiums fit several million.

---

* Charles Babbage's proposed Analytical Engine is not too far removed from what a mechanical-switch-based computer could like.

† The earliest transistor, built at Bell Labs in 1947, actually used germanium. Silicon transistors were easier to mass produce, and by the late 1950s, germanium transistors had already become outmoded.

Watching silicon shrink became a spectator sport. Transistors are at the heart of one of modern electrical engineering's great clichés: Moore's law, which states that the number of transistors on a chip will double every two years or so. Moore's law isn't a law of physics, but a rule of thumb tied to engineering and economics. Moore's law only has authority if chipmakers keep on manufacturing smaller transistors, making more of them for a similar cost.

For decades, Moore's law held firm. By the 1990s, when the smallest transistors had reached channel lengths as short as 250 nm and started to brush against the nano realm's door, some raised the alarm that silicon transistors might stumble after 100 nm. Their worries did little to stop silicon's progress. About thirty years later, the farthest bleeding-edge transistors on the assembly line today measure around 20 nm.* The chips in many home computers now contain hundreds of billions of transistors.

Beneath the surface, in labs around the world, researchers are telling a different story. It's increasingly difficult to ignore the warnings – silicon's progress has begun to slow. Moore's law is teetering. Engineers may have forced silicon too far into the nano realm, and the harder they push, the more spectres they face.

---

* Some chipmakers call these processes '3 nm' or '2 nm', but this is marketing speak; these names haven't correlated to actual sizes since at least the 1990s.

For example, it's hard to manufacture chips down at this scale. Fabricators often 'print' circuitry by shining light upon a wafer, a technique known as photolithography. At first, chipmakers could use visible light. Then, as transistors shrank past the wavelengths of visible light, chipmakers resorted to ultraviolet. The smallest transistors have now blown past all but the shortest-wave ultraviolet, and chipmakers are now considering even-shorter-wave X-rays. Shorter wavelengths bring higher energies, and without proper precautions, energetic X-rays are more likely to completely obliterate your wafer.

Perhaps more damning, silicon itself starts to glitch in the nano realm's depths. Smaller transistors are more prone to overheating for reasons that aren't entirely understood. Even if you could keep them cool, smaller silicon transistors are more prone to leaking current, which somewhat defeats their whole purpose. Plus, silicon has a size limit: it needs three dimensions in order to operate, and there's no known way to make a completely flat silicon transistor. If mandatory 3D seems like a quaint problem now, it won't be if we want to keep making electronics smaller.

If we indeed want to do that – and there's good reason to want that – we'll have to find a successor that can replace silicon in a transistor's channel and operate smoothly in the nano realm. The good news, then, is that scientists have lined up plenty of potential successors. Our candidates range from the mundane and the familiar to the esoteric and outright exotic.

Carbon's champions

When we discussed the history of the atom, we encountered John Dalton and his primitive list of the chemical
elements. For modern chemistry, Dalton's list was only
the start. As chemists catalogued element after element,
first in minerals and later in particle accelerators, the list
sprawled to several times its original size. Today, we count
118 in all, though only around 90 form naturally outside
the lab. A 118-item-long list is too unwieldy for most,
and over the nineteenth century, it gradually gave way to
a table with more structure.

The periodic table comes with a basic organising
principle. Each of its columns generally represents a
different number of outer electrons that an element's
atom can possess. As a result, elements in the same column tend to share certain chemical properties and react
in similar ways. The periodic table's jagged edges thus
frame an elegant map of chemistry's landscape. If we're
seeking a silicon replacement, we can use this map to
guide us.

It might not come as a surprise that some of the most
promising electronic possibilities come from silicon's
upstairs neighbour: carbon.

Once carbon nanotubes entered scientific knowledge
in the early 1990s, researchers were quick to prod at
amazing electronic abilities. In 1998, researchers at Delft
in the Netherlands built the first carbon nanotube transistor by stacking a single-walled tube channel atop silicon
electrodes.

It was the first of many more in the decades since. Some are like the Delft design, made from single tubes; others are made from multiwalled tubes; still others are made from multiple tubes woven together into a thin film. Researchers have been able to create more complex nanotube devices. In 2013, a team at Stanford wired 178 carbon nanotube transistors – with channels made from nanotube bundles – into the first-ever nanotube computer, capable of counting up to 32. In 2019, MIT researchers did them one better by assembling 14,000 nanotube transistors into a 16-bit machine capable of writing words like 'Hello, world! I am RV16XNano, made from CNTs'.

Carbon nanotube electronics wouldn't just be small; they would also be speedy. When carbon nanotubes let electrons flow, the electrons move with efficiencies and speeds that leave silicon in the dust. In theory, a nanotube processor could be ten times as efficient as a silicon equivalent, running thrice as fast on one-third of the energy.

If all this is true, why haven't carbon nanotubes fully replaced silicon yet? The main problem is not one of performance but of manufacturing tubes at high enough quality. Most carbon nanotubes are semiconductors, but recall that not all carbon nanotubes are alike. As we've discussed, if you wrap a carbon nanotube's atoms in certain configurations, you will wind up with a so-called metallic nanotube, which lacks a bandgap and conducts electricity like a metal.

Even a lone metallic tube in a sea of semiconducting tubes is enough to leak current and sully a transistor. We need to manufacture carbon nanotubes such that fewer than one in a million tubes is metallic, and those nano-tubes need to be aligned in the same direction. Today's manufacturing methods are just not good enough, though they are getting better.

What, then, of graphene? Graphene has many of the carbon nanotube's advantages – small size, electron speed – but graphene's problem is that it lacks a band gap. You want your transistor's channel to flow smoothly when the gate is open and to stand still when the gate is closed. Without a band gap, graphene cannot do the latter.

There are a few ways around this. You could pry open a band gap by adding oxygen and nitrogen atoms to the graphene, or you could 'wrinkle' the graphene by denting its structure, but these chemist's crowbars only open narrow band gaps less than half as wide as silicon's. This isn't ideal; the wider your band gap, the higher energies and temperatures your semiconductor can handle. Alternatively, instead of sheets, you could use pasta-like graphene ribbons, which can pick up a much wider band gap akin to a carbon nanotube. Researchers have employed such ribbons as the channels in primitive graphene transistors.

Or you could make a semiconductor by combining a graphene electrode with a solar cell made of a perovskite, already a common substrate for growing the carbon layer. In 2024, researchers at Georgia Tech in the US and Tianjin University in China used this method to create a graphene semiconductor with a band gap closer to half silicon's, which is able to shuttle electrons ten times faster. One day, a semiconductor like this could build complete chips small enough to fit inside corners of the human body that we explored in the last chapter.

2D electronics have some advantages over silicon in addition to size and speed and efficiency. Because carbon nanotubes and graphene are so thin, they let through nearly all the light that strikes them (as we'll see in more detail in the next chapter). This ability could be used for transparent electronics. It's easy to imagine embedding graphene electronics in glass – imagine near-invisible chips mounted in a window or in an ordinary pair of eyeglasses.

Moreover, if you have ever held a circuit board, you know that silicon circuits tend to be rigid by nature. Carbon nanotubes and graphene are, on the other hand, very flexible. Imagine a wafer-thin mobile phone that can roll up or a bendy display woven into your sleeve. We could even build electronics on a sheet of paper – scientists have already shown that it's possible to write electronics onto paper using an inkjet printer. A parcel of the future might come with a printed logic circuit for a label, while a printed medical sensor could sit on the skin like a tattoo.

These tantalising possibilities have driven some researchers, bored of waiting on nanotubes or graphene, to look beyond carbon.

2D imitators

If imitation is the greatest form of flattery, then graphene has no shortage of admirers that scientists have summoned from all across the periodic table. 'Flattery' may be appropriate; 'flat' is the operative word in the world of what scientists call 2D semiconductors.

Take the transition metal dichalcogenides, or TMDs. A TMD contains one atom of a transition metal (an element from the periodic table's metallic midlands) for every two atoms of a chalcogenide (an element that shares a column with oxygen, such as sulphur, selenium or tellurium).* These atoms are lined up in a three-atom-thick structure: two chalcogenide layers sandwiching metal atoms in the middle. Three atoms might not be quite as thin as graphene's sole layer, but TMDs can be treated in much the same way.

The most famous TMD is molybdenum disulphide, or MoS_2. Chemists had long known about MoS_2 in its bulk form. Just as graphene's layers stack into graphite, MoS_2's pyramidal slabs contain layers stuck together by van der Waals forces. Bulk MoS_2 had long served an unglamorous role as a machine lubricant before scientists, in the fashion of Geim and Novoselov's graphene discovery, ripped away flat pieces of MoS_2 in 2011.

This tear elevated MoS_2 from obscurity. 2D semiconductor sommeliers now tend to rate MoS_2 the most capable of the non-carbon candidates. Unlike graphene, MoS_2 does have a natural band gap quite similar to that of

---

* If you're wondering about the other elements in this column, buried in the periodic table's undercity, they're too radioactive for electronics. Polonium is best known as a poison used by Russian spies, and the most stable isotope of livermorium has a half-life measured in milliseconds.

silicon. MoS_2's quality as a transistor is now well-proven. In 2016, researchers at Berkeley combined a MoS_2 channel with a carbon nanotube gate to create a 1-nm-long transistor – attaining the record for the smallest transistor ever built. The next year, researchers in Vienna created a primitive 115-transistor microchip by combining MoS_2 channels with gates of titanium and gold.

The challenge, then, is actually creating enough MoS_2. Ripping sheets of MoS_2 off chunks of its 3D equivalent will give you MoS_2 layers, but better-quality MoS_2 is needed for electronics. 2D MoS_2 can be grown from the bottom up instead, by depositing its atoms onto a substrate, but the process is currently slow. Nonetheless, MoS_2's future looks bright.

Other researchers are seeking semiconductors in the column below nitrogen. The most promising candidate here is phosphorus.* Like carbon, phosphorus can exist in a few different allotropes, each with a different crystal structure. White phosphorus is used for chemical weapons; red phosphorus coats the strike pad on a book of matches. We're actually interested in a third allotrope, black phosphorus, whose bulk form was discovered by accident in 1914 and promptly relegated to the status of a forgotten stepchild.

---

* Here's a good reminder that not all scientists read the periodic table in the same way. People who work with semiconductors tend to call this column 'Group V'. Other chemists call it 'Group 13', saving 'Group 5' for the literal fifth column from the left: the one that starts with vanadium.

Perhaps this was an injustice; black phosphorus is a natural semiconductor, after all. In 2014, several different groups thought to give black phosphorus the graphene-scotch-tape treatment. What emerged from their experiment was a 2D material in a two-tier stepped pattern, which materials scientists quickly named 'phosphorene'. (Like MoS_2, phosphorene today can be grown bottom up on a substrate, using phosphorus atoms from a gas.)

Within the year, a prototype black phosphorus transistor had been forged. Like graphene, black phosphorus lets electrons travel quickly. Unlike graphene, black phosphorus has a band gap, and one that's easy to modify by adding more layers.

However, as with MoS_2, phosphorene is difficult to make. Additionally, phosphorus tends to oxidise when it touches water, even vapour in the air, which tends to degrade any electronics that use it.

The search for new semiconductors has taken us all across the periodic table, but we may be able to find another competitor very near where we started. If carbon and silicon are neighbours on the periodic table, and if they share chemical properties on account of their shared postcode, then it stands to reason that we might find a silicon-based equivalent of graphene. In 2010, a group of French researchers revealed they had grown ribbons of silicon graphene – silicene – upon a substrate of silver.

Graphene and silicene are not actually identical twins. The closer you look, the more differences you will spot. For one, there isn't any known silicon equivalent to graphite (silicite, as it were) that scientists can pull from nature and tear apart. Instead, silicene is an entirely synthetic contestant, almost always grown from the bottom up. Silicene's crystal structure also isn't a perfect flat honeycomb, but rather a field of tilted and misshapen hexagons, resembling a field of sand dunes. Silicene's buckling gives silicene a band gap that graphene's geometrical perfection lacks. Moreover, silicene's band gap is relatively easy to tune by bathing silicene in an electrical field.

In 2015, researchers in Brianza, Italy and at the University of Texas created the first silicene transistor. So far, silicene hasn't received the same attention as MoS_2 or phosphorene, but there are several reasons that may change in future. Researchers expect that silicene is easier than those other contestants to combine with the doped silicon of today's electronics, and silicene can be made bottom-up at much lower temperatures than graphene or other 2D semiconductors.

Silicene isn't a direct mirror to graphene, but that description may fit another candidate. Take the honeycomb structure of graphene, then replace the carbon atoms with alternating atoms of boron and nitrogen. What you've got is hexagonal boron nitride (h-BN). For this reason, h-BN has gained the epithet 'white graphene'.

It must be said that 'white graphene' isn't the most accurate epithet. In this electronic world, hexagonal boron nitride's properties are rather different from its dark counterpart. Boron and nitrogen have different electron layouts, so their bonds don't match up quite as well as those of graphene's carbon atoms. This causes h-BN to mirror graphene in a different way: where graphene is a superb conductor, h-BN is an insulator.

Being an insulator does not preclude h-BN from joining the electronics world. Indeed, h-BN has an important role to serve. Recall that each FET has an insulating layer that separates its gate from its channel. Today, that insulating layer is usually made from silicon dioxide, but engineers can thin it further by replacing it with h-BN.

While h-BN is not graphene's equal, h-BN and graphene can be fast friends. When graphene is layered over a different non-2D substrate, the substrate often contains impurities that disturb graphene's electrons. As a flat material itself, h-BN is far less likely to hold such impurities. When h-BN is topped with a graphene channel, researchers have found that the graphene transmits its electrons an order of magnitude more quickly than with other substrates.

Yet h-BN may not be destined to serve a supporting role forever.

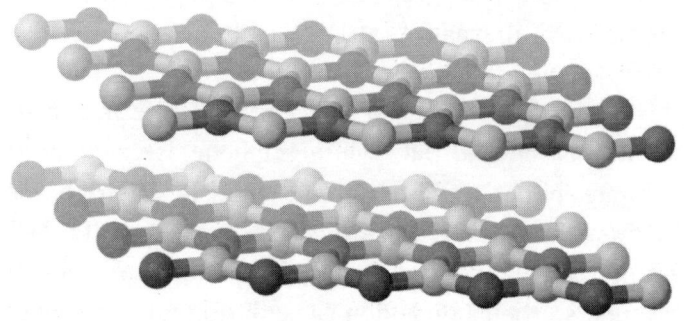

h-BN.

Exotica

In 2024, MIT researchers unveiled a transistor made from two honeycomb-layers of hexagonal boron nitride. But this transistor is less like a silicon transistor and a lot more like a mechanical switch. In the stead of a gate electrode, the upper layer slides one atom over to switch on the transistor, then one atom back to switch it off again.

This transistor is just one of many efforts dedicated to completely restructuring how computers work from very near square one (in fact, around the same time, a Chinese group unveiled a very similar transistor made from MoS_2). These exotic efforts make graphene transistors look positively mundane.

The researchers who made that h-BN-switching transistor believe it could one day be used as a form of computer memory. The RAM in your home computer is what electrical engineers call *volatile memory* – when you switch off your computer's power, the memory vanishes. Mechanical memory is *non-volatile*, meaning that it doesn't need power to keep information in its RAM. Non-volatile memory can offer immense power savings, especially for very large supercomputers.

There are other types of non-volatile memory in progress in the lab. One uses tiny magnetic shards. In addition to electric charge, particles in the quantum world have a property that physicists call 'spin', an inherent angular momentum that might point up or down. When the spins of enough atoms all align, they spawn magnetism. Here is where *spintronic memory* enters the scene.

Spintronic memory proposes that we store information in magnetic domains whose atoms all align either up or down. It's likely to involve the nano realm. In 2008, IBM researchers developed a system that stored information in magnetic bits on a nanowire 'racetrack' and accessed them by shuffling them around the circuit.

Other future computers would completely revamp how computations are done. Many computer scientists think that 2D semiconductors are vital for building a neuromorphic computer, one that mimics a human brain. Rather than the hard logic circuits of computers past, a neuromorphic computer relies on 'artificial neurons' that communicate with each other through electrical signals.

The most sweeping change of all might come in the form of a quantum computer: a device built by physicists who have successfully domesticated the beasts of quantum mechanics. A quantum computer's key component is the qubit. A qubit is where the computer harnesses those quantum creatures to store information and carry out calculations. Where a classical computer deals in absolutes, zero or one, a qubit can be zero and one at the same time. In accessing this quantum maths, a quantum computer could perform calculations in science and cryptography that are utterly impractical for a non-quantum counterpart.

Many objects can serve as qubits, but the largest quantum computers today tend to use superconductors of one shape or another. This sort of qubit must be chilled to fractions of a degree above absolute zero

(-273°C). Qubits are tediously sensitive, and any heat at all can disturb a qubit and throw it into error. If you see an image of a quantum computer's hardware, you probably won't immediately see the computer core that houses the qubits; you'll instead notice the tangle of refrigeration needs required to keep the qubits in that deep freeze. These cryogenics are inconvenient for everybody involved, but they're quite literally the quantum computer's lifeline.

We have already encountered a potential alternative. Quantum dots, as the name might suggest, can function as qubits. Multiple quantum computing groups have shown in the past several years that silicon quantum dots function at -272°C, an entire degree above absolute zero.* While not exactly a tropical holiday, even this single degree drastically cuts a quantum computer's refrigeration equipment.

Quantum dots offer another advantage. Because they're made from semiconductors, it's possible to do something that's significantly more difficult with other sorts of qubit – manufacture qubits in a standard silicon fab. In 2024, Intel engineers demonstrated they could repurpose equipment intended for making conventional computer chips to print qubits instead. If we are ever to mass produce quantum computing hardware, quantum dots may be the way.

---

* It's somewhat telling of what quantum computing is usually like that quantum computer researchers openly speak of these as 'hot dots'.

We may have come full circle by revisiting quantum dots, but we still haven't seen everything that quantum dots can do. In addition to performing at temperatures cooler than the boiling point of helium, they can also benefit from basking in the sunlight.

GOING SMALL TO SAVE THE PLANET

6

At first glance, a solar cell based on quantum dots may not look particularly extraordinary. The pieces emerging from labs have tended to appear less like power sources than printed circuit boards, shaded splotches on a plain background. Yet in the minds of some solar cell makers, these splotches are key to accessing power that standard solar cells cannot.

The core of a typical solar cell is a semiconductor cake with two silicon layers, one n-type and the other p-type. As photons strike the silicon, they energise its electrons, knocking them across the band gap. The electrons flow from the p-type layer to the n-type layer, then out of the solar cell as electrical current. When you see a solar panel on a rooftop or rows of them in a solar farm, this is most likely the invisible process playing out before your eyes. We know that this process works well, and silicon cells are cheap and easy to mass produce, so why bother trying to change it?

In a word, the answer is efficiency. Solar cell developers measure a solar panel's performance in a single number by describing its efficiency – the proportion of incoming light that turns into electricity. A higher-efficiency solar panel can generate more electricity from the same sunlight. The mass-produced silicon solar panels adorning rooftops and solar farms today tend to have efficiencies around 20 per cent. A simple silicon layer cake alone cannot surpass around 33 per cent, wasting two thirds of the incoming energy.

Outpacing that 33 per cent mark – known to solar panel connoisseurs as the Shockley–Queisser limit – is not simple, since it usually requires drastically altering the silicon layer cake. By using tricks like extending the cake with additional semiconducting layers, engineers have been able to push that figure into the mid-40s.

Silicon can only carry solar cells so far. Part of the problem is that the element is best at absorbing light in a limited range of wavelengths, between 500 nm and 1000 nm – roughly between green light and the near infrared. Longer-wavelength infrared simply doesn't carry enough energy to push electrons across silicon's band gap. Bluer light with shorter wavelengths has the opposite problem; it often carries too much energy and gets reflected away. This gives silicon solar panels their bejewelled ocean colour, but in the deep blue is a lost opportunity.

Quantum dots can better reach these blues. By tapping the quantum ability that powers quantum dots, engineers can resize the dots to capture blue wavelengths. Then, by implanting a range of quantum dots in multiple sizes, engineers can make solar cells that better trap the entirety of sunlight's spectrum. Additionally, quantum dots are theoretically better than silicon at squeezing energy from light. When a photon strikes normal silicon, it can only elevate a single electron. When a photon strikes a quantum dot, it can jolt multiple electrons.

The patches on a quantum dot solar cell, then, are quantum dots contained within splotches of a colloid.

When scientists created the first quantum dot solar cells around 2011, their prototypes had efficiencies of around 4 per cent. As of this writing, just over a decade later, that figure has grown to around 18 per cent. It's still not a match for today's solar cells, but progress has been rapid, and scientists have plenty of motivation to keep pushing forwards. In theory, according to calculations done by US National Renewable Energy Laboratory researchers, quantum dot solar cells could reach efficiencies of around 66 per cent.

Quantum dot solar cells are shining examples of how we can use nanotechnology to create a more sustainable future. One day, not so far away, quantum dot solar cells might adorn the rooftops and balconies of a yet-unbuilt city. What other kinds of nanotechnology might we find in such a city?

Plastic solar cells

We might find other kinds of solar cells. Silicon solar cells of today are thick and rigid, slabs of semiconductor encased in metal and glass. Our future city's solar cells might resemble sheets of paper instead: thin, feather-light, easy to bend and easy to stick on a wall.

These might be organic solar panels, which eschew the silicon for a plastic-like polymer ('organic' simply means that the polymers are carbon-based molecules). When plastic solar cells first emerged in the 1990s, many relied on a mix of plastic and fullerenes, those carbon spheres whose discovery helped kick off nanotechnology in the first place. Here, the plastic plays the part of semi-conductor, catching sunlight to excite electrons. Electrons in plastic don't travel very far, but solar cell crafters can pair the plastic with fullerenes, which excel at wicking electrons away.

Flexible fullerene solar cells, though still rare, have already been installed on places like building walls and curved wind turbine shafts. Meanwhile, some newer organic solar cells find a different place for a different form of carbon. That's in great part due to graphene (or carbon nanotubes), which can easily act as a solar cell's electrode.

For electricity to actually flow out of a solar cell, it must be placed in a complete electric circuit. Solar cell makers usually do this by sandwiching the solar cell between a pair of electrodes, one of which rests atop the solar cell – look at a standard silicon solar cell, and you'll probably see a metallic electrode lattice adorning its face. Gold and silver are common choices thanks to their electrical conductivity. Though a jeweller's metal can make a solar cell glossier, that gloss is opaque and blocks valuable surface area where photons could otherwise strike the solar cell.

Paperweight organic solar cells instead rely on more transparent electrode materials, often tin oxide. Graphene can outperform tin oxide. Where tin oxide at best lets through around 90 per cent of light, graphene can outmatch it with transparencies upwards of 97 per cent. We don't know how to shrink tin oxide to thicknesses of a nanometre, but a nanometre is easy for a graphene electrode. Organic solar cells with graphene have now broken the 15 per cent efficiency mark.

If that isn't enough, an alternative may be on the way. Today's materials scientists also get quite excited by the possibility of combining graphene with a perovskite.* Perovskites are materials with a cage-shaped crystal structure, and they can be shaped into films that can go down to thicknesses of 500 nm.

Perovskites have become the rising stars of the solar cell world due to how rapidly they've advanced. The first perovskite cells of the early 2010s had efficiencies around 3 per cent. Just over a decade later, perovskites have already leap-frogged many of their competitors.† The record for a flexible perovskite solar cell today is around

---

* They're also quite excited by the possibility of combining perovskite with fullerenes, for the same reason plastic solar cell makers like fullerenes – fullerenes help boost electrons on their way out of the perovskite.

† Researchers have reached around 29 per cent by combining a perovskite layer with a silicon layer.

25 per cent.* The record for a flexible cell with graphene isn't quite there yet, but it's pushing 20 per cent.

Ultrathin, lightweight, flexible, near-transparent solar cells could be ubiquitous in our city of the future. Our city's parks might host greenhouses with solar panels in their roofs; ordinary homes might hold solar panels in their windows, unseen by most passersby. Inside our city's buildings, wireless gadgets and household appliances might come with solar cells that draw electricity from sunlight during the day and from artificial lamps at night.

They'll likely need accompanying batteries.

Electric clothing

Rechargeable batteries have been around since the nineteenth century, but when the lithium-ion battery first entered everyday life in the 1990s, it held more energy and lasted longer than the old-style lead-acid and nickel-based batteries that came before. The lithium-ion battery has now become nearly as integral to twenty-first-century life as the silicon transistor.

Today's lithium-ion batteries come in many shapes and sizes, but they all rely on a common design: a positively charged cathode and a negatively charged anode,

---

* There is another problem for the moment, which is that most perovskites contain lead. This limits their widespread use, but there's a lot of research into finding less toxic combinations.

separated by a solvent. When we switch on a phone or drive an electric car, lithium ions flow from the cathode to the anode, generating electric current. When the battery is plugged in to recharge, the ions flow in reverse and pile back up at the anode. Then the battery is unplugged, and the cycle begins again.

The whole process depends on how well the anode and cathode can hold onto the lithium ions; the more ions that can land, the higher the battery's capacity. Today, most lithium-ion batteries use anodes made from graphite. Many experimental anodes use nanoplatelets of graphene instead, which have more surface area for ions to use. Replacing graphite with graphene can increase a lithium-ion battery's capacity by as much as 30 per cent. Graphene also doesn't degrade as quickly as graphite does, and a three-year-old mobile phone with a graphene battery could run for longer than a week on a single charge.

Our city's buses may run on graphene batteries, having become as mundane as the batteries in today's electric cars. Our city's denizens may also be very familiar with an energy competitor now lurking in graphene's shadow: the MXene.

Though they may sound like extraterrestrial parasites, MXenes (pronounced 'max-enes') are actually rather more harmless. MXenes are a category of 2D materials comprised of two metal layers sandwiching a middle layer of oxygen and carbon or nitrogen. These atoms are fitted into an undulating accordion-shaped structure. An MXene's folds give an MXene sheet even more surface area than one of graphene, providing more space for ions to rest.

The first MXene was synthesised in 2011, and unsurprisingly, MXenes quickly gained a reputation for their energy storage capabilities. There's also a lot of ongoing research into using MXenes to create batteries that don't use lithium at all: batteries that use sodium ions instead; hot batteries that shuttle ions between molten sodium and molten sulphur; and iron-air batteries that discharge by essentially allowing their iron to rust, then revert to clean iron as they recharge.

For some devices, our city might not use batteries, but capacitors. Where a battery stores energy within chemicals, a capacitor holds it as an electric field. Capacitors 'charge up' far more quickly than a battery, but we don't use capacitors to power devices today, because they also discharge far more quickly than a battery. A bog-standard capacitor also just can't store enough energy to power a phone.

However, researchers are stacking layers of graphene, MXenes and other flat materials like MoS_2 into so-called supercapacitors. We're a long way from using supercapacitors in the real world, but a successful supercapacitor could combine the best of two worlds – the charging speed of a capacitor and the capacity of a battery.

On a clear day, it's quite possible that a flat-dweller in our city might look out their windows and down upon masses of pedestrians clad in solar-panel-embedded jackets. Our flat-dweller might watch one of those pedestrians tap a handheld electronic device on their jacket's cuff. That handheld, a barely recognisable descendant of our mobile phone, might contain an ultrathin supercapacitor. A full charge, one that lasts a week of heavy use, could take mere seconds.

Windows into energy savings

The flat-dweller might have watched all of this through a colour-changing window. A building's west-facing window could automatically shade late in the day, reducing the nuisance glare from a low-angled sun (not to mention the energy needed to cool a room in direct sunlight). These windows could become as standard as double glazing is today. Indeed, one estimate suggests that they could reduce household energy use by up to 40 per cent.

So, it's not as surprising as it may first seem that many researchers are looking into smart windows. A smart window may contain a thin metal film of copper or bismuth that deposits itself on the glass when an electric field is activated, then dissipates after the field is removed. Or, windows might rely on a nanometres-thin film of tungsten trioxide, a colour-changing semiconductor. Tungsten trioxide is what scientists call an electrochromic material – its colour changes at the touch of an applied voltage, shifting from transparent to a shaded blue or black.

We might find other examples of nanotechnology in our city's architecture. For the same energy-saving reason, the city's windows and walls might come insulated with an aerogel.* Take silica gel, like the sort that comes in 'do not eat' packets to keep packaged food dry, and strip away

---

* An aerogel once held the world record for the least dense solid known to exist. That record now belongs to the mystical aerographene, a recently created material that has a similar structure to aerogel but replaces the silica with narrow graphene spindles.

all of the gel's liquid. On the nano scale, what remains is a tangled bush of thin silica strands separated by air-filled nanopockets. Each pocket may be tiny, but combined, they fill over 90 per cent of the aerogel's volume. To the naked eye, the result may look like a slab of frozen smoke, and since some aerogels are less dense than air, an aerogel block might feel like smoke if you lift it.

Because aerogel is so sparse, heat is barely able to crawl through its structure. Heat one side of a centimetre-thick aerogel blanket, and you can touch the other with your bare hand. Aerogel has been used in spacesuits for precisely this reason. Aerogel's sparse structure also makes it quite soundproof, so our city's night owls won't have to worry about upsetting their neighbours. We've actually known about aerogels for about a century now, and aerogel insulation is already on the market today, though it is significantly costlier than traditional insulation.

Our city's walls might quietly contain other sorts of nanoparticles. Concrete is made by blending cement with water and rocky slurry, and researchers have already tried adding various nanoparticles to the mix. Carbon nanotubes make concrete more resistant to cracking; iron oxide nanoparticles make it more resistant to abrasion and stronger under compression; titanium dioxide nanoparticles make it better able to stay clean. Our city's builders may add graphene to the mix, too, to make concrete that's even stronger and more impermeable to water.

The concrete in our flat-dweller's building might last far longer than the concrete we recognise, and if there ever comes a time when this building must be demolished, the concrete block might remain sturdy enough to be reused in a construction project elsewhere. Today, concrete is the world's most used material, and concrete alone is responsible for nearly a tenth of the world's CO_2 emissions – we've good reason to reuse it.

If we travel to one of our city's industrial districts, we might find many more sorts of sustainability-boosting nanomaterials. Our city's chemical plants may rely on reactions that are sped up by nanoparticle catalysts. Our city's factories might coat their machine parts with lubricants that contain graphene or fullerenes. By reducing friction, these sorts of nanomaterials can reduce the amount of energy that escapes into a material's surroundings as useless heat.

Lubrication and catalysis may seem like chemistry's equivalent of streamlining office workflows to improve employee productivity, but a lot of energy is actually wasted in friction.

Cleaner living through nanomaterials

Let's stay in this industrial district for a moment longer. There, far from most city-dwellers' eyes and minds, we might find a towering apparatus of fans that suck air into a large drum-shaped vessel. Inside, the air passes through sheets of a nanomaterial that separates out the carbon dioxide and stores it elsewhere.

Today, carbon capture is an unproven technology with a track record of underdelivering. Much of carbon capture's support comes from the fossil fuel industry, which has not done wonders for the technology's credibility. Part of why carbon capture has not caught on, so to speak, is that existing methods are simply too expensive. New types of nanomaterial could change that.

The people who built our city's carbon capture plant might thank nanoscientists like Omar Yaghi, now at Berkeley. Yaghi caught his first glimpses of the nano realm as a child in Jordan, leafing through the pages of chemistry books. Searching for a way into this strange world, Yaghi moved to the US aged fifteen with a limited command of English and a dream of becoming a chemist.

Yaghi would become the world's first specialist in 'reticular chemistry', the craft of taking small molecules and elegantly stitching them into large frameworks such as MOFs (metal-organic frameworks).* Chemists had observed frameworks like these since the 1960s, but in 1995, Yaghi and colleagues became the first to actually synthesise one – creating patterns worthy of an ornate rug, molecular flowers arranged with gaping holes in between.

Although a MOF's molecular structure sprawls far beyond the nano realm, it is made from nano-sized components, and it demonstrates the nano realm's affinity for surface area. A MOF has very nearly the highest surface-area-to-volume ratio of any material in existence. One teaspoon of a MOF may contain the surface area of

---

* There's a naturally occurring analogue of reticular chemistry in the form of minerals called zeolites. In fact, zeolites are used for many of the same applications as MOFs and COFs.

two tennis courts. Recall, then, that chemistry is a game of surface area. The more surface area you can work with, the more reaction you can perform.

MOFs have loads of possible applications ranging from detecting poisons to catalysing industrial reactions, but their potential to trap greenhouse gas has captured the most attention. Even if a MOF looks like a rug from its molecular structure, it acts more like a sponge in the real world. Just as a washing-up sponge takes up water by storing it within a pockmarked structure, MOF nanocrystals

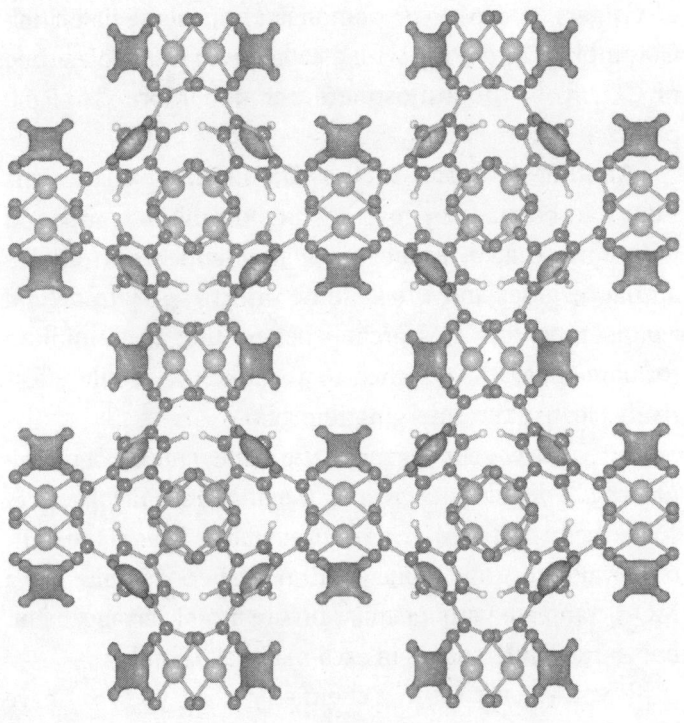

A MOF.

take up carbon dioxide from the atmosphere and stow it within the MOF's gaps. Hypothetically, you could place a MOF out in the open air, or you could fit one into a flue to scrub the emerging exhaust gases.

Since the 1990s, chemists around the world have synthesised thousands of different MOFs by tweaking the atoms that form this molecular rug's petals and branches. Several pilot projects have already started up with MOFs at their command. One of the more promising MOFs for carbon capture is called CALF-20, built with a core of zinc. CALF-20 was developed by chemists at the University of Calgary. A CALF-20 demonstration plant in British Columbia is operating with a capacity to remove a tonne of CO_2 from the atmosphere per day. More CALF-20 plants are on the way.

Another promising MOF is aluminium formate. This MOF can be made by reacting an aluminium compound with formic acid, a substance that appears in apples and aubergines and which some insects spray to defend against predators. Researchers believe that an aluminium formate apparatus attached to a smokestack could effectively cleanse the emerging flue gas.

Alternatively, our city might use a different sort of molecular rug. A decade after creating the first MOF, Yaghi and his colleagues gave the MOF a younger sibling. This is the COF, or 'covalent organic framework'. A COF looks quite like a MOF, complete with a similar ornate atomic arrangement, but the key difference is in each material's bonds.

If you've ever sat in a chemistry class, you may have learned that atoms form different types of bonds. MOFs are built on ionic bonds, which form when one atom steals an electron from another atom; the thief becomes a negatively charged ion and its victim a positively charged one, and the two stick together. COFs instead contain covalent bonds, which form when two atoms share electrons. Covalent bonds are what we find in carbon nanotubes and graphene. A COF is usually more difficult to create than a MOF but tends to be more lightweight and more stable.

As MOFs are cycled through capturing and releasing CO_2, they tend to degrade quickly. COFs, then, may address that problem. One of the newest COFs to emerge from Yaghi's lab is the earthy-yellow-coloured COF-999, unveiled in 2024. With the human eye, it's easy to confuse a vial of COF-999 for a jar of turmeric, but turmeric doesn't have molecular tendrils to grab CO_2 from the air.

MOFs and COFs aren't the only nanomaterials that can scrub carbon dioxide from the atmosphere. Some researchers are trying to capture carbon with graphene sculptures or MXene sheets, while others have proposed sprinkling seed-like metal nanoparticles that encourage CO_2-absorbing algae to bloom about them. None of these techniques will likely be a panacea for global warming, but if the materials become cheap enough, they could do a lot to mitigate future climate change.

CO_2 is not the only greenhouse gas in the sky, nor is it the only pollutant that engineers are trying to corral with nanomaterials. There's a lot of interest in using graphene filters to cleanse methane, a potent greenhouse gas. A recently launched pilot plant in Cambridgeshire accomplishes this by turning methane into graphene; the plant separates each methane molecule's hydrogen and carbon atoms, saves the hydrogen for fuel and forms the carbon into graphene pellets.

The very walls of our city's buildings may help clear the air, too. We can already find a prototype of how this might work if we look to a particular building in Milan's northwest suburbs. The Palazzo Italia was first built as the host country's pavilion for Milan's Expo 2015. From the outside, the pavilion looks like a plate-glass prism shrouded in cement gossamer. The cement contains a secret ingredient: titanium dioxide.

We've already seen how titanium dioxide can make hardier concrete, but it's also a photocatalyst – when light strikes the material, the titanium dioxide is excited into reacting with its surroundings. In open air, those surroundings are smog-forming air pollutants. The titanium dioxide reacts those pollutants away into inert, harmless salts. The Palazzo Italia was clearly a showpiece, and its titanium dioxide didn't come in the form of nanoparticles, but materials scientists at TU Wien in Austria recently created a titanium-dioxide-nanoparticle-laced paint that effectively cleaned both itself and the surrounding air – at least in the lab.

Filters of the future

Our future city's taps may flow with water that's passed through a nanomaterial. Some of our city's residents might actually harvest their water right from the atmosphere.

Researchers today have shown that this works with nano-materials that have considerable surface area: a MOF, for example. In experiments so far, MOFs work best when the weather is hot and humid, leaving the air filled with water droplets that can latch onto the MOF's surface.

If our city is near enough to the coast, we might get our water from a desalination plant, but it probably won't be like desalination today. At present, the chief method of turning salty ocean water to drinkable fresh-water is reverse osmosis, in which pressure is applied to force water across a membrane that does not let the salt pass. The science behind reverse osmosis is elementary enough for schoolchildren to learn, but actually placing reverse osmosis into action is both expensive and energy demanding.

As early as the 1960s, researchers started developing 'nanofiltration', which used membranes pockmarked with pores measured in nanometres. Nanofiltration is usually used in a multistep process alongside reverse osmosis, but there's evidence that nanofilters can stand on their own. A late 2010s study conducted in arid Qatar found that one nanofiltration method was just as effective as reverse osmosis but consumed nearly a third less energy.

Materials have of course advanced a great deal since the 1960s, and more recent desalination experiments have taken advantage of newer 2D materials. Researchers have made membranes from graphene, MoS_2 and boron nitride alike, or sometimes a combination of the three. Perhaps the most intriguing material is graphene oxide – a sheet of graphene with hydrogen and oxygen atoms attached.

These dangling atoms are especially good at grabbing the ions that float in water. Engineers can tune them to pick out specific ions corresponding to specific pollutants.

Some scientists have also proposed creating membranes whose pores are carbon nanotubes, with diameters as small as 0.8 nm. These are small enough to prevent salt from passing, yet large enough to allow an unimpeded flow of water. Interestingly, this structure mimics cell features known as aquaporins, little channels in a cell's membrane that allow a cell to transport water between itself and its surroundings.

We shouldn't forget the other end of our city's water supply. The city's sewage treatment works might rely on similar membranes to separate water from its contaminants. Just as graphene oxide can pluck out ions from seawater, they can pluck out unpleasant contaminants from wastewater. Our city might couple such filtration with a dash of silver nanoparticles. Silver ions are well-known bacteria-killers that silver nanoparticles produce in abundance, and silver nanoparticles combine that with a surface area advantage.

There are cases when our city might need a less passive sort of cleanup. Suppose an oil tanker – perhaps, from the perspective of our city's denizens, a relic of a past era – catches fire and bursts its tanks within sight of our city's walls. What is to be done?

Historically, cleanup crews rushed to disasters like the *Exxon Valdez* and Deepwater Horizon by treating petroleum-soaked waters with chemical mélanges called dispersants. These work by breaking large oil pools down into smaller droplets, which are easier to disperse and easier for microbes to degrade. It's a distinctly imperfect process. Dispersants are prone to pushing oil into even more dangerous places, such as soft and sensitive coastlines.

A cleanup crew from our future city might instead deploy nanorods or carbon nanotubes that can pick up oil particles with much less collateral damage. In 2012, an international team first tested a sponge-like amalgam of carbon nanotubes designed specifically for this purpose. The sponge is both hydrophobic – repelling water – and oleophilic – attracting oil into its structure. It could absorb more than 100 times its own weight in oil. After the oil is squeezed out, this team's sponge can be reused.

In the future, cleanup crews could do even better – they might pick up individual pollutant particulates from the water with the aid of nanorobots. In fact, scientists have already tested precisely one such system.

A nanobot cleanup crew

If you want to make a nanobot, you must first choose an appropriate material. These nanobots' creators chose nanoparticles of magnetite, a mineral comprised of iron and oxygen. Magnetite is, as its name suggests, magnetic, a property that is crucial to the nanobot's operation.

The magnetite is made by pouring ammonia into a solution of iron and chloride. After stirring the mixture, heating it to near-boiling, then cooling it back to room temperature, iron starts to coalesce with oxygen atoms to form magnetite nanoparticles from the bottom up. Each nanoparticle grows until reaching a diameter of about 200 nm. Each is then outfitted with tiny arms of a temperature-sensitive polymer. As these arms sweep across pollutants, they chemically capture them.

Because the magnetite nanoparticles are magnetic, they can be controlled by the twitching of a magnetic field from the outside. This magnetic control system turns these elaborate nanoparticles into genuine (if very basic) nanorobots. By shifting the magnetic field, a human can easily manipulate a swarm of nanorobots through a liquid.

In one experiment, the nanobot-makers tested their newfound creations on a solution filled with arsenic. It's estimated that some 140 million people around the world consistently drink water with arsenic levels above the World Health Organization's recommended limits; arsenic is linked to skin damage, heart disease and cancers. In this experiment, after the nanobots had a go for less than two hours, they had reduced the arsenic level by more than half.

These nanobots need no fuel. Their creators weren't the first to craft water-cleaning nanobots, but other bots of their kind relied on chemical reactions to propel themselves. Instead, these ones move to the dance of a magnetic field. Like the nanoscalpels we saw in chapter 4, they can be steered by the magnetic touch of researchers on the outside, manipulating this field.

Moreover, nanobots like this can easily be reused. Once they've finished cleaning up one pool of liquid, they can easily be sprinkled into another pool, placed under a different magnetic field and employed to do their cleanup task yet again.

This is the nanorobot swarm of today, as developed and tested in 2022 at the University of Chemistry and Technology Prague. They are ingenious tools that take advantage of the abilities the nano realm has on offer. More importantly, they can be made inexpensively and quickly. If they can be manufactured for cheap enough, it's very possible that the wastewater treatment plant of a future city might install a large-scale magnetic field and use it to sweep nanobots across their quarry.

Thanks to their magnetic control, these nanobots resemble their counterparts that we've already seen: those brain-tumour-busting nanoscalpels which medical researchers could likewise control with a magnetic field. But ask the public to imagine 'nanobots', or 'nanotechnology' as a whole, and their mind probably does not go to nanoscalpels and nanoparticles with rudimentary limbs – but elaborate machines on a tiny scale.

NANO NIGHTMARES 7

The latest James Bond film at the time of writing, *No Time to Die*, brought nanotechnology into the world of international spycraft – after a fashion.

For Bond and for MI6, the problems begin in a laboratory high inside a Central London skyscraper. Within its glass walls, scientists have developed a swarm of nanobots whose purpose is to kill. A wielder can programme the nanobots for assassination with the judicious application of a target's DNA. These nanobots can spread from person to person like a virus, out of sight and out of mind until it is too late for the hapless victim. Naturally, the laboratory falls victim to an explosive break-in, and the situation quickly spirals out of hand.

Unless somebody, somewhere has discovered ways of creating nanomachinery and nanobots that are decades more advanced than anything known to the scientific community, these events are not possible to play out with today's science. Nonetheless, the writers of *No Time to Die* were certainly not the first to imagine nanotechnology gone berserk.

Take the *Metal Gear Solid* franchise of video games, which infamously used 'nanomachines' as an explanation for giving its characters preternatural powers. Soldiers in the video games' universe are implanted with nano-machines that boost their stamina, purify the water they drink and unlock their gun like a mobile phone's finger-print reader. Nanomachines bestow some of their bearers with the power to heal from near-fatal injuries and the power to transform into a musclebound beast.

Then there's 2002's *Prey*, one of the later novels writ-ten by Michael Crichton of *Jurassic Park* fame. Released at the height of the original nanotechnology craze, it depicts the foibles of a company that has hijacked an *E. coli* bac-terium to manufacture swarms of nanobots. One swarm gains the ability to replicate and evolve – and quickly sets about stalking the engineers who brought it to life.

Back on the screen, there's 'The New Breed', an epi-sode of the 1990s *Outer Limits* reboot. A scientist creates a type of nanobot that can heal virtually any ailment in the human body. One of his colleagues, diagnosed with termi-nal cancer, injects himself with the untested nanobots in a last-ditch effort to get his old life back. The nanobots work. Unfortunately, they work too well. They start correcting 'flaws' in the poor man's body, 'improving' his flesh into horrors he did not desire.

It is not difficult to understand why nanotechnology elicits so many nightmares. The nano realm overlaps with the familiar bulk world, but the nano world is unseen.

'Nanotechnology' offers viewers a world where invisible horrors can creep out of nowhere. It is interesting that few of these visions actually seem to focus on the materials involved, the materials which are key to making any of this nanotechnology work in the first place.

People who are more familiar with nanotechnology have dreamt up their own nano nightmares. Few of those visions can hold a sleepwalker's candle to the apocalyptic one devised by one of nanotechnology's figureheads himself. K. Eric Drexler dedicated a whole section of *Engines of Creation* to the possibility that nanotechnology might, very literally, consume the world.

After Drexler imagined molecular assemblers (see page 51), he imagined a scenario in which those assemblers escape our control. He suggested it might start with, say, a simple malfunction in an assembler's programming. It incites a nanomachine to start making copies of itself. Each of those copies starts making more copies, and so on. If they're small enough, and if they're capable enough, they could rapidly form a deadly swarm – a patch of nanomachinery that spreads despite humanity's best efforts. In just a few hours, a single nanomachine seed might be able to sprout into a behemoth weighing more than a tonne.

Even if we did have the weapons to fight back, we might not have the time: should we destroy one piece of the machine swarm, a single survivor could rebuild it with frightening speed and inhuman efficiency. These nanomachines could transform their surroundings into energy far more efficiently than even the hardiest plant; they could consume biological matter far more efficiently than any animal or any known life-form. The nanomachine swarm might take no longer than two days to spread across the whole of the Earth, consuming everything in its way – buildings, railways, trees, blades of grass.

There would be nowhere left for humans to live, nothing left for humans to live on. There would be nothing left at all for most life on Earth. All that remained of our planet's surface would be a hostile nanomachine shell: as Drexler termed it, a 'grey goo'.

At the moment, there is little reason to worry about grey goo. We do not have self-replicating nanomachines that can hijack human bodies or consume the environment. Nonetheless, writers aren't the only people who have thought about the possibility that nanotechnology could kill.

Nanotech goes to war

Every now and again, scientists gather to evaluate the sorts of existential threats that might menace humanity in our near future. They read the technological entrails to predict the chances that billions of people will die from nuclear war, super-potent bioweapons, out-of-control AI and the like. At one such gathering, in Oxford in 2008, the convened scientists placed 'molecular nanotech weapons' near the top of their red list.

Digging into the details does not make for an optimistic read. The survey estimated that nanoweapons had a 25 per cent chance of killing at least a million people, a 10 per cent chance of killing a billion and a 5 per cent chance of causing outright human extinction by 2100. That last figure was on par with the extinction risk from superintelligent AI; it was greater than that from horror pandemics or nuclear wars.

Predicting human extinction is a terribly inexact science, if it can be called a science at all, and its results no doubt change with time.* But those scientists were not the first to concern themselves with nano-weaponry, nor would they be the last.

In Cambridge, Massachusetts, across the Atlantic, near the heart of MIT, an unassuming office building between biotech firms and coding bootcamps is home to ISN, the Institute for Soldier Nanotechnologies. For more than twenty years now, the US Army has joined forces with MIT nanoresearchers.

Many uses of nanotechnology in the armed forces are fairly mundane and designed to save lives rather than destroy them. ISN's research goals range from better armour to emergency medicine. Soldiers need not have a monopoly on this sort of research. It's easy to imagine these inventions taken onto the job by firefighters, repair technicians or other civilians who work in hazardous environments. Indeed, many of the same applications are worked on in labs across the world by nanotechnologists who have no connection to the military whatsoever.

---

* For example, you might imagine that a survey taken at the height of the ChatGPT craze fifteen years later would say somewhat more about superintelligent AI.

The fact remains that, from the time nanotechnology emerged into the public sphere around the 1990s, many have sounded alarms that this sort of research could, hypothetically, create nanoweapons designed to kill. They imagine that some of the nanotechnologies we've seen might become double-edged swords in the wrong hands. Gene editing could engineer better bioweapons. Chemistry techniques from the nano realm could create more potent poisons. Nanoelectronic components could underpin electronic warfare on a very tiny scale. Some analysts around the turn of the millennium even warned that terrorists might get their hands on nanotechnology. They imagined 'mini-nukes' that employ nano-sized components to stuff a hundred-tonne warhead into a handbag.

There has never been a proven nano-attack from anyone. Yet whispers of death have followed nanotechnology for decades. After Venezuelan president Hugo Chavez died in 2013, Russian state media published unsubstantiated rumours that Chavez's long-time opponents in the US government had assassinated him with some kind of cancer-causing nanoweapon. You can find books breathlessly warning that a nanoweapons arms race has already begun between the world's most powerful nations.

We can't say whether such an arms race is actually playing out, but we can say that nanotechnology is a matter of international intrigue. Take the case of Charles Lieber, a longtime Harvard University professor of nanoscience. For decades, Lieber's group had led daring expeditions into the nano realm, specialising in building and laying nanowires. His laboratory's creations included nanowires woven into logic circuits and nanowires which could record electrical signals from live cells.

Lieber's career took a dramatic turn on one cold January morning in 2020 when US law enforcement appeared at his doorstep. Lieber, they announced, had accepted funding from a Chinese government programme without proper disclosure. At the same time, the FBI issued arrest warrants for two Chinese nationals, one of them a Chinese military officer whom US officials accused of joining Lieber's lab under false pretences.

Lieber's case made international news and drew protests from the scientific community, but if you ask many other scientists who work in the nano realm about what keeps them awake at night, you'll probably get a different answer from geopolitics.

When nanomaterials go wrong

In 1982, a group of Belgian researchers tested whether a polymer called polyalkylcyanoacrylate could damage living cells. They were not trying to create weapons. On the contrary, they were trying to shape polyalkylcyanoacrylate into nanoparticles for delivering medicine, something we've discussed and something that's since become commonplace. Determining the toxicity of a material destined to enter the human body, then, was a logical thing to do.

At concentrations the scientists expected to use in the clinic, the polymer was safe. At far higher concentrations, however, the researchers noted signs of cell damage.

In one sense, the group's results deemed their nanoparticles unlikely to cause harm if they were used for their intended purpose. In a completely different sense, their paper was the first confirmation that, if nanoparticles could help heal the body, they could also help harm it. That Belgian group unintentionally became the founders of nanotoxicology: the study of how the nano realm can poison the human body.

Nanomedicine walks hand-in-hand with nanotoxicology. The two fields may seem antithetical, but they are really studying two sides of the same problems. Helpful nanoparticles can enter the body using the very same pathways as noxious nanotoxins. A nanomedicine gone awry can easily become a nanotoxin.

If nanotechnology is all about harnessing the fact that nanomaterials behave differently from their bulk counterparts, nanotoxicology tells us that nano-incarnations of a particular material affect the body in different ways. Many abilities from the nano realm can turn into double-edged swords. Nanoparticles' greater surface-area-to-size ratio means more potential for harmful ones to latch onto tissue. Even deadlier is a nanoparticle's ability to enter corners of the human body that are otherwise too small to enter. We try to insert nanomaterials in the human body for good, but if the wrong nanomaterials enter the wrong places, the effects could be disastrous.

As an example, take the quantum dots we discussed in chapter 5. Scientists are very interested in using quantum dots as a means of imaging inside the human body, but they're also aware that many quantum dots contain cadmium, a heavy metal linked to kidney and bone damage and cancer. Quite a lot of research is ongoing into the toxicity of quantum dots, and it's given us evidence that quantum dots can accumulate in organs.

Nanoparticles might not enter the body via a carefully planned injection. They can also enter the body through the respiratory system, through the oesophagus or by passing through skin – especially if the skin is burned or broken. Nanoparticles are also small enough to enter the bloodstream. If we want medical nanoparticles in the bloodstream, that's certainly a good thing. If we don't, that can have debilitating consequences.

This is why nanotoxicology is important. Nanotoxicology tells us that even if we know a material's toxicity at the bulk scale, we cannot assume that it will behave the same at smaller sizes. The human body simply does not respond to particles in the nano realm in the same manner. Nanoparticles smaller than around 30 nm tend to pass through biological tissue and infiltrate cells in ways that their larger counterparts simply do not.

Take silver nanoparticles. They're common catalysts in industrial plants, and they're often used medicinally as antimicrobials and as drug couriers. Many bioengineers are working on weaponising them against cancer cells. But small enough silver nanoparticles can accidentally wind up inside perfectly healthy cells. Inside, they can start to break down and release microbe-killing silver ions in the wrong places.

Although we've said that the human immune system is good at recognising things that do not belong, the human body's natural defences are not designed to counter unrecognised particles this small. Nanoparticles can be effective endocrine disruptors, interfering with the system that the body uses to distribute hormones. A particular concern is the possibility that certain nanoparticles might break through the brain-blood barrier. Here, again, nanoparticles are double-edged swords. If drug-delivering nanoparticles or magnetic nanoscalpels can enter the brain to purge cancer cells, then so might toxic nanoshards of heavy metal to scourge otherwise healthy brain cells. There's already some evidence that metal nanoparticles like silver and gold can build up in the brain.

As a result, it's up to nanotoxicology to examine nanoparticles as their own threat. One of the better-scrutinised such particles is titanium dioxide. Nanoparticles of titanium dioxide are found in sunblock, concrete, ceramics, paints, food colourings, window coatings and many more places. They're safe on skin – applying titanium dioxide sunscreen is surely safer than letting your skin bake in ionising radiation – but some animal experiments have hinted that titanium dioxide nanoparticles can cause genetic damage inside the body. It's a controversial subject, but European Union authorities saw enough reason to ban titanium dioxide nanoparticles from food.

A nanoparticle's shape also can influence its toxicity. We can look to a ghost of our present to find out how. Asbestos is one of the modern world's most fearsome killer materials, with a human death toll in the hundreds of thousands at least. This building material is so deadly because it fragments into fibres that can scar lung tissue. Asbestos fibres are typically about the size of bacteria, but some can get even smaller, as short as 100 nm. Carbon nanotubes are fibrous, too, and it's logical to wonder if they might have the same effect. The evidence is still inconclusive, but some studies have indicated that carbon nanotubes can irritate skin cells in mice and damage human lung cells.

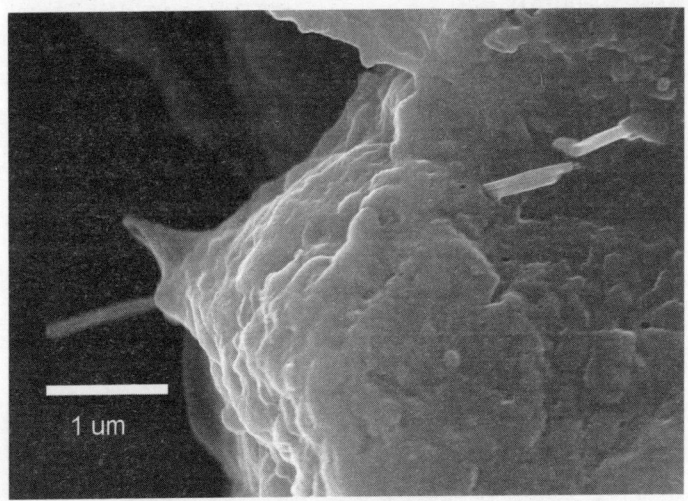

1 um

Carbon nanotubes penetrating a lung cell.

In the end, why does it matter that nanoparticles are dangerous? Could we not simply limit our exposure to them and ensure our safety? Unfortunately, the solution isn't quite that simple.

Nanomaterials gone amok

Look at your mobile phone's weather app or at today's forecast. Next to your city's temperature and cloud cover and precipitation, you might also see an air quality index. If that index has enough detail, you might see counts for particulate matter (PM) emissions: specifically, PM10 (measuring particles smaller than 10 micrometres, or 10,000 nm in diameter) and PM2.5 (measuring particles smaller than 2.5 micrometres, or 2,500 nm).

Particulates can arise from many different sources. When things burn, their smoke tends to contain all sorts of noxious aerosols. Road vehicles kick up dust and shed bits of rubber as their tyres roll down the pavement. Plastic that we don't properly recycle will fragment into microplastics. Taken into the human body, particulates like these can irritate the eyes, choke the lungs, damage DNA. There is very good reason why weather forecasters track particulates.

The good news is the developed world has made a lot of progress on curbing air pollution, and so too have particulate levels plummeted. In the UK, PM10 and PM2.5 levels have both dropped by half between 1993 and 2023.

But PM pollution is still a threat to those who are exposed to it, as you'll know well if you've ever lived downwind from a wildfire or spent time in a smog-choked city. The bad news for us, then, is something that you've probably worked out from those particulate sizes: nanoparticles are, by definition, also particulates.

Some environmental scientists have coined the label 'ultrafine particles' for a class of particulates smaller than PM2.5. Their definition of 'ultrafine particle' counts particles in the environment with a diameter of less than 100 nm – most nanoparticles, then. Scientists still don't agree on how best to count and classify ultrafine particles: they debate whether we should do so based on size, mass, chemical composition or some combination of the above.

This is a problem. The globe already moves well over a million metric tonnes of nanomaterial every year, expected to reach 3.5 million tonnes by the early 2030s. A great many of those nanomaterials wind up in consumer goods, and those nanomaterials will stay with their consumer goods to landfill. When they degrade, they might spread nanoparticles into land, sea and air.

3.5 million tonnes per year may only be a very tiny tip of a much larger nanopollutant iceberg. Scientists now believe that intentionally created ultrafine particles, 'engineered nanoparticles', are far outnumbered by 'incidental nanoparticles', which enter the environment as unintended byproducts of other processes. Counts of nanomaterials in the environment have listed substances like silica (used intentionally in medicine, but more often unintentionally kicked into the air by dust storms), carbon particulates (produced from fires and from burning fossil fuels) and nanoplastics (even tinier shards that break down from microplastics).

These ultrafine particles often aren't wonder materials. They're often ordinary pollutants that have broken down into more annoying and possibly more dangerous sizes. Much like bulkier PM emissions, they can cause respiratory health issues, heart disease and genetic damage, according to studies that have examined the matter.

Since ultrafine particles are lighter than other types of particulates, they can stay in the air longer than their larger counterparts. They can enter the soil and infiltrate plant life. They can also wind up in Earth's oceans and rivers, where they can endanger aquatic organisms or dissolve into water. Again, the effects of these phenomena are still not well-understood.

The technologies we discussed in the last chapter can certainly help us identify some of these ultrafine particles and cleanse them from the atmosphere. But if we're trying to find nanopollutants, they are not always easy to identify, especially if they're made of carbon-based materials. Organic, carbon-based nanoparticles are like needles in a carbon-based haystack.

Many of the same technologies that help the environment can easily become deadly with subtle changes in how they're used. A growing number of researchers are trying to plant nanomaterials into the soil to help grow better crops. Sprinkling nanoparticles into fertiliser, for example, can make crops more resistant to high salt levels, but scientists in Tanzania have determined that sprinkling the very same nanoparticles in doses too high can be toxic.

Laying down the law

What is to be done?

There are existing laws that govern the use of nanomaterials and nanotechnology. In the medical world, getting a work of nanotechnology from the lab to the real world is subject to regulations already designed to ensure physicians and patients can have confidence their treatments will be effective and safe. Medical interventions must clear a gauntlet of clinical trials before they can reach general use. Nanomedicine is absolutely no exception.

Those gauntlets have already caught certain nanomaterials in their grasp. In 2008, regulators denied approving one nanomaterial – a certain type of iron oxide particle intended for use as a contrast agent during MRI scans – after some trial subjects experienced severe muscle pains and 'allergic reactions which resulted in one death'.

For nanomaterials outside medicine, however, the situation isn't as straightforward. Many countries regulate the materials from which nanoparticles are fashioned, such as heavy metals, but, again, scientists are increasingly well aware that nano-versions of certain materials can act quite differently from their bulk counterparts.

Some regulators are now mindful of this. US government workplace safety monitors have set limits for how many titanium dioxide and silver nanoparticles employees can breathe in on the job. The EU has penned guidelines on the safe and sustainable use of 'advanced materials', as nanoparticles are sometimes called.

Part of the problem with setting regulations around nanotechnology is that nanoparticles are so diverse in both size and substance. You can make nanoparticles out of virtually anything, after all, and how large can a nanoparticle become before it ceases being a nanoparticle? It's difficult to make rules that cover such a wide range of materials.

One point where there is agreement is that we must better understand how nanomaterials behave and what they can do, both in the human body and in the environment. Nanotoxicology and environmental nanoscience are the fields of examining these nano-nuisances before they become full-blown nano-nightmares, and they've grown in popularity in the last several years.

How can we study the toxicity of a certain nanoparticle without actually forcing humans to ingest pinches of it? Scientists can rely on cell models grown on their lab benches, which allow them to simulate the effects of exposing tissue to nanoparticles. Additionally, as we've seen, there are an increasing number of studies that intentionally expose lab rodents to nanoparticles. Since a rodent's anatomy isn't actually very different from a human's, lab mice are useful stand-ins for studying how nanoparticles (or ultrafine particles) might impact humans.

Nanotoxicologists can also play detective. For example, certain workplaces, like factories that manufacture nanoparticle-containing sunscreen and other nano-inclusive products, tend to expose their workers to more nanoparticles. Researchers can study these workers' health to determine if and how nanoparticles have affected them.

If measuring nanoparticles in the human body is difficult, then measuring nanoparticles that are floating about in the air is utterly tedious. A great part of the task is simply taking each ultrafine particle's measurements. Even if we do capture them, traditional means of sizing them don't work. Scientists usually size up larger particulates by shining them with visible light and measuring how they scatter it, but the wavelengths of light mean that we can't do this with particles smaller than about 300 nm.

Fortunately, environmental nanoscientists have developed several tools to help in this task. They might use a condensation particle counter, which attracts particles into a vapour-filled chamber; inside, each particle becomes the seed of a droplet that grows until a laser beam can bounce off its surface and count it. They might use an electrostatic classifier, which catches particles and exposes them to an electric field; nanoparticles of different sizes will arc through the field along different paths. These tools let scientists view particles as small as 0.4 nm, just a few atoms across.

Beyond merely observing, environmental nanoscientists can also run experiments that test what nanoparticles do to their environment. It isn't exactly ethical science to intentionally release nanoparticles into the wilderness, but scientists can use proxies here, too.

These proxies are known as *mesocosms*.* A mesocosm is an artificially curated biome within a controlled environment, like a wetland in a greenhouse, or a freshwater pond in an aquarium tank, complete with enough organisms to create a functioning ecosystem. After environmental nanoscientists create a mesocosm, they can intentionally release nanoparticles within it. They can then observe how the nanoparticles cycle through their terrarium, how soil absorbs the nanoparticles and how the nanoparticles affect plant growth.

The mesocosm has been around for decades, and its uses are hardly limited to the nano realm, but it's an appropriate tool for studying nanomaterials. If nanotechnology as a whole is a bridge between the world around us and the quantum world, then the mesocosm is a bridge between tiny lab benchtop experiments and the larger, uncontrollable natural world. As nanomaterials grow increasingly abundant in all sorts of applications, this sort of work will only grow in importance.

Studying toxins and pollution is an important aspect of understanding the nano realm. Some environmental nanoscientists enter the field after having started their

---

* You can actually find directions online to make your own mesocosm in a glass jar.

careers in other parts of nanoscience. Pollution may not be a desirable part of life at the bottom, but it is an inevitable one. Learning what exactly it is and how exactly it works are some of the first steps towards cleaning it up. Moreover, engineers are constantly creating new nano-materials or adding new sorts of nanoparticles to existing materials. We won't know how new materials affect us until we test them.

Furthermore, if we can succeed in managing these nano-headaches, it will be all the easier to build some of the more aspirational technologies that nanoscientists are developing today – including more capable nanobots.

NANO DREAMS

8

Robots in the blood

If fiction imagines nanobots as hidden killers within the body, then science imagines them as something else entirely. We've already seen two examples: nanoscalpels that can kill cancer cells and nanoparticles whose arms pick pollutants from water.

These concepts may be ingenious, but they are also quite basic. We can only barely describe them as machines. Nonetheless, they have a crucial characteristic that separates them from simpler nanoparticles: a human operator can control them. A nanoparticle can certainly be dressed up with proteins in the hopes of tricking a cell to invite in the nanoparticle, but once that nanoparticle enters a body, its journey is entirely at the mercy of bodily functions. Nanobots have other ways of moving.

There's no commonly agreed definition of a nanobot, and not everyone would call our examples nanobots. (A magnetically steered nanoparticle is a crude analogue to a remote-controlled drone, and not everyone would call a drone a robot.) Yet these nanobots are performing tasks upon matter in the nano realm, and they can be controlled from the outside. For their creators, this is enough to earn the nanobot label.

Magnets are only one way to propel and control a nanobot. Another possibility is ultrasound. In 2012, researchers at the University of California, San Diego, created traffic-cone-shaped pellets filled with droplets of fluorocarbon. When the researchers pulsed the cones with ultrasound, they ignited the fluorocarbon and launched the cones as fast-moving 'microbullets'. Doctors of the future may be able to shoot similar bullets as kinetic weapons into cancer cells with the power of sound.

A nanobot could also obtain energy from the body itself. Some researchers have proposed using the oxygen and glucose that is plentiful in human blood to power a robot through the same chemical reactions that cells rely upon. In 2014, researchers at Berkeley fed zinc-coated polymer tubes to lab mice, each morsel about 20,000 nm in length and stuffed with gold nanoparticles. As a tube descended into the mouse's stomach, the zinc reacted with the surrounding stomach acid, forming bubbles that propelled the tube into the stomach lining. As gold nanoparticles are common couriers for drug delivery, it's easy to imagine this method used for treatment.

The biological world is already teeming with other robot-like mechanisms. Many microbes propel themselves through their watery world using a propeller-shaped organelle called a flagellum. We might create nanobots of our own that mimic this corkscrew. In 2015, chemists from Rice University and North Carolina State University created a nanocraft with an artificial flagellum. Their vehicle was a single 244-atom molecule shaped like a Formula 1 steering wheel, with a flagellum-like propeller out back. When they activated the propeller by casting it in ultraviolet light, their craft sped through a solution at a breezy 2.5 centimetres per second.

What could you do with a drone inside the bloodstream? As with nanoparticles, nanobots have the potential to be useful as scouts within the body. If we could outfit them with sufficiently small electronics, we could equip one with a 'lab-on-a-chip' that can test for anything from acidity to temperature to the presence of certain chemicals.*

As with basic nanoparticles, you could deliver drugs. In 2024, researchers from Edinburgh and Shanghai showed off a cluster of magnetic nanobots, each about one-twentieth the size of a red blood cell. The researchers inserted a swarm containing billions of bots into a rabbit's artery and magnetically guided them into an aneurysm-stricken brain. Once there, the researchers used the same magnets to heat the nanobots to their melting point. As each nanobot melted, it released its payload of blood-clotting agents.

---

* We might be able to fit tiny transmitters on these robots as well. In 2007, researchers at Berkeley wired carbon nanotubes into a functional radio antenna. They played the world's first nanoradio track, succeeding in picking up a broadcast of Eric Clapton's 'Layla' transmitted from across the room.

If these are what nanobots look like today, what might they look like in a generation or three? In 1959, during his talk in California, Richard Feynman related a thought that his former graduate student Albert Hibbs had shared with him.* Hibbs had imagined that a cardiovascular patient might swallow a 'mechanical surgeon', which would then enter the patient's blood vessel and operate on the heart with a tiny blade.

The classic image of a 'mechanical surgeon' may be the image of *Fantastic Voyage* (see page 57), a vessel that can navigate the human body to a cell and perform surgery on the spot. We're unaware of any labs trying to shrink humans to blood-cell sizes, but many researchers are working to build automated vessels just like the film's one. Much in that film's spirit, nanotechnologists call their idea a 'nanosubmarine'.

Like so many aspects of nanotechnology, this is one field of research that has barely begun. Physicians are decades away from regularly deploying nanosubmarines on search-and-destroy missions to play out *The Hunt for Red October* amidst red blood cells. Nonetheless, the interest is there, and with continued research, the technology might follow.

Meanwhile, one of Feynman's other predictions may have begun coming true.

---

* Though Hibbs did little work in the nano realm, he did have a prolific career as a designer of satellites for NASA's Jet Propulsion Laboratory.

Molecular machinery

The factory of a nanotechnologist's dreams may start with a spare part called the catenane.

Imagine two ring-shaped molecules intertwined as two links in a chain, such that the chain cannot be broken without snipping one of its rings. Chemists named this strange piece of molecule-work after the Latin for chain, *catena*. A molecular chain might seem like the chemist's idea of a child's finger trap, but its development was the seed of yet another Nobel Prize.

There had been sporadic attempts at forging a catenane from the 1950s, but it was a French team at Strasbourg University led by Jean-Pierre Sauvage who finally did it in 1983. Sauvage's group succeeded in

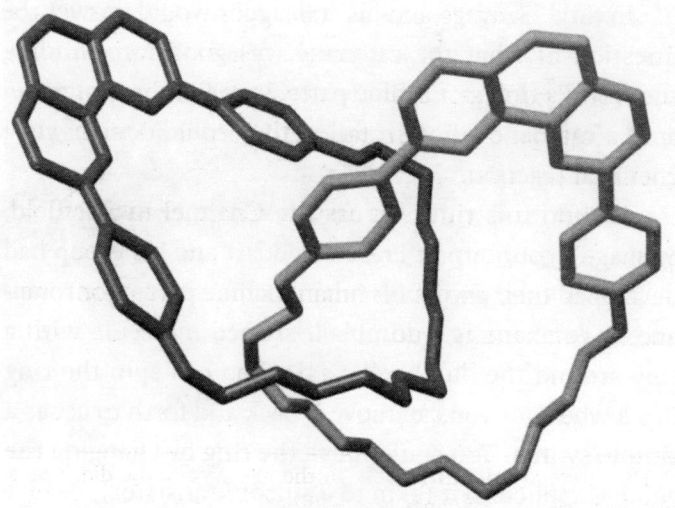

A catenane.

creating a seemingly paradoxical molecule by starting with a smaller ring-shaped molecule, inserting a second crescent-shaped molecule through the ring's hole with the help of a copper ion, then closing the crescent into a second ring. In the years since, researchers have used methods like this to create an assortment of more complex catenanes: molecular crosses, molecular Celtic knots and six-pointed molecular Borromean rings.

Chemists didn't know what the catenane was actually good for, but they did know that it wasn't any ordinary molecule. Catenanes belong to a field called 'topological chemistry'. When we think of a molecule, we usually think of atoms joined by traditional ionic or covalent bonds. Topological chemistry instead handles molecules connected by interlinked loops and joints, not unlike how metal parts in the human-sized world join together to create machines.

In time, Sauvage and his colleagues would answer the question of what the catenane was good for: building the world's tiniest machine parts. In 1994, the group created a catenane whose rotation they could control via a chemical reaction.

Around this time, across the Channel in Sheffield, Sauvage's counterpart Fraser Stoddart and his group had developed their own molecular machine part – the rotaxane. A rotaxane is a dumbbell-shaped molecule with a ring around the dumbbell's axle. You can spin the ring like a wheel, or you can move it back and forth to act as a simple switch. You could move the ring by changing the voltage applied to it (akin to a silicon transistor), or by a chemical signal, or even by shining light on the rotaxane.

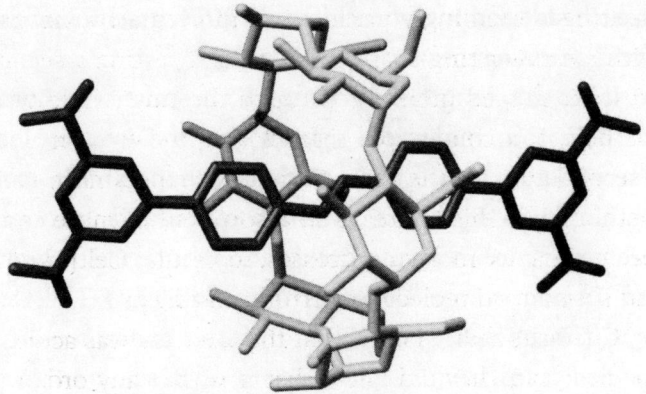

The rotaxane.

Building machine parts is the first step in constructing a molecular device. It's an entirely different task to actually assemble them into a fully functional molecular machine. Enter a third group, led by Ben Feringa at the University of Groningen. In 1999, Feringa's group interlocked two crescent-shaped molecules – two molecular 'rotor blades' – into a molecular ratchet. When Feringa and his colleagues lit up the molecule in ultraviolet, the ratchet rotated in one direction. The group had created the world's first molecular motor.

These discoveries heralded more molecular machines of increasingly sophisticated sorts – molecular pistons, molecular windmills, rotaxanes that could flex and contract like molecular muscles, rotaxanes assembled into a 'molecular elevator' that could rise and fall on command by a whole 0.7 nm. Feringa's group designed a bone-shaped central molecule with four rotating molecular motors, two on each side and two at each end, in 2011, assembling the world's first nanocar.

It wasn't long before nano-motorsport was born. In 2017, four molecules started the first-ever molecular endurance race on a gold-atom nanocircuit atop a tabletop in Toulouse. (Though billed as a 'nanocar race', the molecules actually lacked internal motors and instead ran on electrons from an STM.) The victor, a triskelion-shaped 'nano dragster' constructed by a Swiss team from the University of Basel, traversed 133 nm.* Lest nanocar racing become a historical footnote, Toulouse hosted a second nano race in 2022; its farthest-travelled car was a Japanese entrant that moved an entire 1,054 nm.

Richard Feynman had foreseen this; decades earlier, he had envisioned a very tiny automobile (see page 38). Feynman had also envisioned that tiny machines could one day synthesise chemicals and move atoms as we desire. When chemistry labs began piecing together molecular machines, many observers believed that Feynman's ideas of nano-industry were within reach.

Assuming that molecular machines really are that capable, then how much farther could we go? K. Eric Drexler, who helped popularise nanotechnology, believed that the answer is quite far indeed. As we've mentioned

---

* This is about one-trillionth the length of the 1894 Paris-to-Rouen, often considered the world's first motor race. At those distances, nano racing is less like Le Mans and more like a science-fiction arcade racing game. Van der Waals forces from the circuit's barriers push and pull at cars shaped like *Star Wars* podracers.

(see page 51), Drexler dreamt of a 'molecular assembler' that could carefully guide atoms into place, in whichever order and shape we wished. In dreams, such a machine could assemble virtually anything at all – imagine the replicator from *Star Trek* that can generate food and household objects on demand.

It bears mention that Drexler's ideas have long courted controversy.* Many lifelong nanoscientists take a chilly view of his work. In this matter, his loudest critic was Richard Smalley – one of the buckyball's co-discoverers. Smalley, who had devoted his post-buckyball career to the carbon nanotube, believed that a true molecular assembler was impossible for two chief reasons. First was what Smalley called the 'fat fingers problem' – he thought that even an arm the size of an atom wouldn't be precise enough to control other individual atoms. Second was what he termed the 'sticky fingers problem' – he felt an atom-sized arm would invariably rub against and react with its surrounding atoms in unpredictable, uncontrolled ways.

We may not have to wait all that long to learn whether Drexler or Smalley was right, because primitive molecular assemblers already exist.

In 2013, researchers at the University of Manchester developed a molecular machine that could grab amino acids

---

* A particular non-nanotechnology idea that has irked some of Drexler's peers is Drexler's support for cryonics, the scientifically unrecognised idea of freezing the dead such that future humans can resurrect them.

from a preloaded track and assemble them into a predetermined protein. In 2017, the same group did one better, creating a molecular machine that could synthesise two different molecules; the researchers could switch the assembler between the two by adding or removing a hydrogen ion. These Manchester efforts are very far from a replicator, but synthesising molecules is one of the hardest parts of chemistry today – if you want to make large quantities of a drug, for instance, even these rudiments can be exciting.

There's a lot of excitement over the future of this extremely new science. Three of those early molecular machinists – Jean-Pierre Sauvage of the catenane, Fraser Stoddart of the rotaxane and Ben Feringa of the molecular motor – earned the 2016 Nobel Prize in Chemistry for their work. The prize committee's announcement boldly compared molecular machines to the bulk steam-powered machines that had catalysed the Industrial Revolution two centuries earlier.

Closer to the present, molecular machines have plenty of more modest uses. As with many other types of nanotechnology, they could carry drugs into the human body – imagine building motorised boxes, directing them to the proper corners, then shining light upon them, activating the motors, opening the boxes, and unleashing the drugs inside. Molecular machines are also already being used in computer parts. In 2007, Stoddart and other researchers assembled rotaxanes into a rudimentary chip of mechanical memory, the size of a white blood cell, with 20 kilobytes of RAM.

This last use raises a question we've yet to answer: how small, exactly, can you make a computer?

Computers beyond the nanometre

We've seen ultrasmall and ultrathin transistors made from carbon nanotubes or 2D semiconductors, but as exotic as their materials might seem, most are still made from familiar architectures. The electrical engineer of today would likely recognise them as transistors. Yet it's possible to build even smaller electronics, even deeper down within the murkiest depths of the nano realm, by building them from even smaller objects – individual molecules.

A computer made from molecules would look utterly alien. The neat layers of a traditional transistor might give way to convoluted bond structures. Bits of information might be stored on single atoms. Too small for any sort of traditional chipmaking technique to matter, entire circuits might grow bottom-up inside a chemist's beaker.

The molecular computer is also a concept with a deeply chequered history. Its story is one of a scientific idea that has repeatedly failed to die. The idea dates back to a time when the silicon transistor was new, when computer components were large enough to see with the naked eye. In that context, building a computer circuit from molecules was a miniaturisation shortcut. It was a way to bypass the entire ladder of Moore's law, the entire zoom into the nano realm, and to skip right to the bottom.

Around 1960, some researchers tried 'growing' molecular circuits from germanium atoms, but they never quite managed it. In the 1970s, a New York University graduate student named Arieh Aviram imagined a world in which circuits could be assembled from molecules. Aviram and his supervisor Mark Ratner proposed a starting point: a 'molecular rectifier' that could transform two-way alternating current to one-way direct current. In a typical circuit, you'd need a sizeable electronic component to manage this, especially with 1970s technology.

Aviram and Ratner instead imagined trying to do it with a molecule. Such a molecule would operate like a semiconductor's p-n junction – stack p-doped silicon atop n-doped silicon, and electrons will only flow in one direction. Rather than fussing about with semiconductor junctions, Aviram proposed accomplishing it with a certain type of covalent bond (see page 115).

In a familiar story, molecular electronics did not really take off. The tentative forays of Aviram and his successors were overshadowed by the silicon juggernaut, which by the 1970s was well on its way to revolutionising the world. Silicon's progress made the idea of molecular electronics seem hopelessly quaint. Yet, molecular electronics did not die out completely. A handful of labs around the world continued to press on. By the 1990s, some of them had actually succeeded in building Aviram's puny rectifier. Then, a few years later, molecular electronics exploded onto the world stage from a place we've briefly visited: Bell Labs.

Bell Labs by the turn of the millennium still hosted an eye-watering amount of cutting-edge research, but its best days were in the past. Its parent company had been broken up by a court decision in the 1980s; the laboratory itself had changed hands several times before winding up in the hands of a Nokia subsidiary. Its situation turned even more precarious after the dot-com bubble burst and sank the telecoms industry with it. Budget cuts loomed. Bell Labs needed any financial life raft it could find. Its researchers came under growing pressure to find results that could be turned around for revenue. A computing revolution would certainly help.

It was around this time that a modest Bell Labs intern-turned-staffer named Jan Hendrik Schön claimed to have discovered a transistor with a single-molecule channel. Schön's molecule was a carbon-based one similar to a plastic. Schön's work promised to herald the next great computing revolution, the beginning of a shift from silicon computers to carbon-based plastic crystals. Alas, what Schön had actually done was fake his data on an unprecedented scale.

Schön's fraud made global headlines and broke into perhaps the early twenty-first century's greatest scientific scandal. If many outside the field have heard of the molecular transistor, it is thanks to the fallout. This episode cast a shadow that still hangs over the field to this day. Many molecular electronics researchers, from seasoned professors to nascent graduate students, left the field for less tainted pastures. This wasn't entirely a bad thing; many of the castaways found themselves working on other technologies, and it's possible that the Schön affair actually boosted the nascent study of graphene.

Still, and again, the idea of a molecular computer simply refused to die off. In 2002 and 2003, even as investigators combed over Schön's fraudulent work, several other labs announced they had indeed created single-molecule transistors (some of which used buckyballs for channels). The earliest ones required temperatures near absolute zero, but room temperature counterparts followed by 2005.

A fully-fledged molecular computer is still in the future, but we do have many of its building blocks. Scientists have been able to craft an assortment of molecular logic gates. IBM researchers have discovered a way to store zeroes and ones in individual atoms of iron and the rare-earth element holmium. There are now numerous examples of single-molecule LEDs. The arrival of rotaxenes and the coming of age for carbon nanotubes and graphene have given researchers new toys. One possibility is that engineers could build atoms-thin circuits by laying molecules atop a substrate of graphene.

If molecules haven't formed functioning computers yet, they have already led to some of the world's tiniest gadgets. In 2022, one biotech company announced it had planted individual molecules onto a silicon chip to fashion a molecular sensor. Their sensor could detect all sorts of biological reactions and read the bases in a DNA sequence.

That brings us to another type of data storage. Far removed from Schön's transistors, biology may have already delivered us a molecular computer.

Biological computers

Inside a cell, as we've seen, it is the molecular machine known as the ribosome which reads instructions from strands of mRNA and applies them by assembling amino acids into protein. When Manchester researchers developed a molecular machine that could assemble amino acids into a protein, what they had actually done was create a very crude copy of a ribosome's output function.

In 2022, the same Manchester lab showed off a 'tape-reading molecular ratchet'. This machine featured a very tiny version of a VHS player's read head, sliding down a molecular tape. As the head rose and fell over bumps on the tape, it 'read' out those contours as bits of information, like the zeroes and ones of a computer. A major difference was that each bit on the tape had three possibilities: zero, one or minus one.*

The molecular ratchet of Manchester takes care of a ribosome's input function. Soon, something like it might be combined with the group's amino-acid-assembler, or something like this, to create a functional artificial ribosome. Imagine feeding your artificial ribosome with a customised reel of molecular tape to programme the molecules you want manufactured.

The Manchester group's efforts at copying the ribosome are another example of how nanoscientists like to mimic the nanotechnology that already exists in the biological world. In the same vein, nature has already given us a sort of nano-information-technology: DNA.

Many of us already tend to think of DNA as biological computer code, a thread of programming instructions for

---

* This sort of computer is known as a ternary computer. Some early computers, especially in the Soviet Union, actually did use ternary logic.

the biological machinescape. DNA even comes with an in-built data storage format that one-ups the three-base molecular ratchet. Cells can store data as one of four possibilities: the four genetic bases, As, Gs, Ts and Cs. Given all this, what stops us from actually using DNA to compute?

This was the question a Californian computer scientist named Leonard M. Adleman asked himself in the early 1990s.* Adleman had a go at using DNA to solve what mathematicians call the travelling salesman problem. Imagine plotting seven points on a graph, representing seven cities on a map. Adleman wanted to find the shortest route between the seven that passes through each city only once.

Adleman encoded each city and intercity path as a particular DNA sequence, then swirled numerous copies of each sequence together in a test tube. Cities and paths joined into DNA routes. Adleman sifted through the DNA combinations, and whenever he spotted an invalid DNA sequence – a route that visited the same city twice, for example – he chemically reacted it out of existence. Adleman kept at this work until his tube only contained valid routes. The shortest remaining route was the solution.

This problem, more formally known as the seven-point Hamiltonian path problem, is elementary for a silicon computer (or, for that matter, a human mathematician with

---

* Adleman is probably far better-known as the man who coined the term 'computer virus'.

pen and paper). Adleman spent an entire week plucking at DNA to solve a problem that a normal computer could have crunched in seconds. Moreover, Adleman's 'computer' could only run one programme; a multifunction DNA computer would not exist for another 25 years. Nonetheless, Adleman's work kick-started a new sort of computer science. Today, we've got DNA noughts-and-crosses programmes and DNA neural networks.

Those who work with DNA computers have reason to believe that genetic code will be the computer code of the future. Unlike many forms of standard computers (but like the non-volatile memory we briefly encountered in chapter 5), DNA doesn't need to be powered on in any recognisable sense. If we want to encode information into DNA as Adleman did with cities and flight paths, then DNA's four bases are arguably much better at storing large amounts of information than a standard computer's electronic bits. A single gram of DNA can store some 700 terabytes of data, enough to fill an entire bookshelf's worth of hard drives at home in the 2020s.

What's more, as we've said previously, we already have the tools to write and rewrite DNA. We can control the biological machinery that nature evolved to perform these tasks, and we can print DNA sequences in any order we please.

In 2023, chemists in Shanghai showed off a DNA integrated circuit. They programmed their circuit with a set of around 2,000 short genetic snippets. The circuit's core was a set of DNA-based logic gates, augmented with a DNA register, a device that tells the computer which data is needed where. When the DNA performed a computation, the circuit transferred that information and 'wrote' it into the register.

Where Adleman's test tube 'computer' randomly combined DNA strands that needed manual sifting, a DNA register can automate that away. The Shanghai researchers managed to perform computations ranging from calculating square roots to identifying genetic molecules.

DNA computers are also in their infancy. One of their current hurdles is that they are still extremely slow by comparison with electronics; the Shanghai computer is certainly faster than Adleman's, but it still required hours to perform basic calculations. Another problem is that, even with the register, the Shanghai computer needs constant human supervision.

If bio-computer scientists can overcome DNA's hurdles and continue to create more sophisticated DNA hardware, scientists could programme biological circuits in a number of interesting ways. In the near future, such circuits could make for super-sensitive diagnosis tools or drug delivery vehicles coded to deliver precise amounts of a drug into cells at precise locations.

Folding DNA

As a physical molecule rather than electrons in storage, DNA can act as something that computer code cannot: construction material.

As yet, we haven't discussed DNA as a nanomaterial, but there's no reason we should not. It's remarkably durable. Single-stranded DNA is quite flexible, allowing it to be bent into contorted shapes, which stiffen when the second strand is added to form a complete double helix. DNA conducts electricity well enough that researchers have already spun it into wires. It also hardly needs mentioning that DNA is perfectly compatible with all life on Earth.

Scientists knew how to fold DNA more than a decade before Adleman conjured a rudimentary computer in a test tube. The credit belongs to a late biochemist named Nadrian Seeman. In around 1980, Seeman was sitting in a pub when his mind's eye flashed to the image of a six-pointed fish from an M.C. Escher woodcut. Seeman realised biologists already had the tools to bend DNA into that strange shape. When one strand of a DNA double-helix is longer than another, the longer strand juts out into an awkward 'sticky end' that's easy to manipulate. Seeman could grip onto a DNA strand's sticky end, move it about and tie it with other strands.

Yet sticky ends are even more powerful than this because of how DNA's base pairs work. DNA is composed of Cs, Gs, As and Ts; but in a double helix, Cs always bind to Gs and As always bind to Ts. When these pairs are in separate strands, as is the case in a sticky end, they tend to attract each other. DNA-benders can use this attraction to their advantage. By carefully coding strands with the right base pairs, they can 'programme' DNA to fold and assemble itself into very precise shapes. A common method of doing this is to code a long 'scaffold' strand, which attracts other strands that staple onto the scaffold in a pre-planned order.

The result is that DNA can be folded into a bewildering variety of shapes – flat lattices, polyhedra, hollow boxes, vase-shaped ellipsoids.* What can do you do with this DNA origami?

It's an artistic medium, for one. In the lab, nano-sculptors have created DNA smiley faces, crude DNA maps and a DNA study of the *Mona Lisa*. Less whimsically, DNA origami can also form the basis of molecular hardware – the DNA register we previously saw, for instance – or sensors for detecting viral RNA. Some researchers have proposed we use it to lay the keel of a DNA nanosubmarine.

Scientists can also wrap DNA into neat packages for carrying drugs. Fold DNA into an octahedral nanocage, and the result is a faux-virus that can sneak drugs past a body's immune system. Researchers at Karolinska Institute in Sweden successfully folded DNA into a 'kill switch'. The environment around cancer cells tends to be more acidic. When the DNA parcel approaches the tumour, the lower pH level triggers its kill switch. The origami unfurls and a cancer-killing amino acid polygon emerges.

---

* The crowning achievement of DNA origami might be 2020's 'meta-DNA', perhaps the biomolecular equivalent of three goblins wearing a trench coat to pass as one average-sized human. Meta-DNA, or M-DNA, consists of six strands of DNA fused together to form a gigantic double-helix.

More complex DNA creations are certainly possible, too. In 2000, while Jan Hendrik Schön walked the corridors of Bell Labs, a group of researchers elsewhere in the office complex tied three strands of DNA into a pair of 'molecular tweezers', which could be opened and closed by applying additional spare DNA strands. In the decades since, DNA-smiths have created DNA motors, DNA rotors and 'DNA walkers' that can creep along a DNA track. More sophisticated DNA walkers can even partake in 'DNA assembly lines', picking up cargo of DNA snippets and combining them as they go.

Researchers have shown that DNA machines can carry non-DNA molecules, too. Just as DNA holds the code for assembling proteins within a cell, we could one day programme our own DNA to assemble customised molecules without.

As DNA may present the most promising form of molecular computer today, so too may it present the most promising sort of molecular machinery.

Feynman imagined nanomachinery in very mechanical terms, like the tools he might have found had he visited an automobile factory. He imagined nanocars and nano-lathes. DNA nanotechnology shows that, at the current pace of research, we might be bound for a future where our nanomachines take after biology instead. As we've seen, nature is already resplendent with nanomachines.

Biology, then, may represent the future of the very bottom. But what of something closer to the top – the first idea we encountered?

Reaching for the sky

It's ironic that the space elevator is actually one of the older ideas in this book; the dream of climbing to orbit predates nanotechnology as a modern field by several decades. It was the pioneering Russian rocket scientist Konstantin Tsiolkovsky who first imagined something like a space elevator in the 1890s.

More than a century and a quarter later – despite the progress in nanomaterials that we've seen, and the progress that seems inevitable in the decades to come – the space elevator is admittedly still one of the more speculative ideas powered by the nano realm.

That does not mean the space elevator is taken as a joke. Over the years, it has been the object of NASA workshops and detailed astronautical studies. Space elevator construction is a serious field that has attracted serious engineers and serious engineering proposals. It seems that the idea is simply too compelling to let go.

Today, there are quite a few who have hopes of constructing one. The Obayashi Corporation, one of Japan's largest construction firms, proclaimed a wish to build a space elevator by the year 2050. There are whispers of a Chinese effort to build one by 2045. A handful of startups have sprouted in recent years with promises of their own.* The world appears to have plenty of willpower, but if any

---

* At least one startup promising a space elevator has admitted its real goal was to sell carbon nanotubes.

of these proposals are serious, detailed plans have yet to see the public eye.

The real hurdles come in the form of some major technical issues that remain unresolved. The largest is the cable itself. Graphene and carbon nanotubes should theoretically have the required strength, but our manufacturing methods don't yet produce them at the quality that's required – or the size, for the largest unbroken nanotubes and graphene sheets we've made can be measured in centimetres. Some proposals call for composite materials that mesh carbon nanotubes with plastic or carbon fibres, but these are works in progress.

Even if we did find the material tomorrow, we haven't got the tools to build it. How will we synthesise enough material to build a 100,000-kilometre-long cable? How will we practically lay a gigantic cable in the sky, where the pull of Earth's gravity shifts drastically from the cable's bottom to its top? A space elevator might promise easy travel to orbit, but how do we put all its heavy infrastructure up there in the first place?

Again, there are quite a few who would try. Many of the puzzle pieces needed to make a space elevator have already begun to snap into place. Take a mission known as STARS-EC. In 2021, researchers from Japan's Shizuoka University tested a method of laying a mock elevator cable between two satellites. Though their tether was made of steel and only stretched about 22 metres, this test showed the world – or, rather, the few engineers paying attention – it was possible to lay a cable in space.

We might one day view this little-heralded STARS-EC experiment as one of the key steps that showed building a space elevator was possible. In the meantime, there can be no doubt that the necessary materials will continue improving.

ZOOMING OUT

9

In the course of this book, we've seen many of the ways that nanotechnology shapes the world today, from quantum dots to mRNA vaccines. We've seen many of the ways that researchers are shaping nanotechnology in the lab. We've also seen many of the more fantastical ways that nanotechnology's dreamers have imagined nanotechnology might shape the world.

Considering those three things, where does nanotechnology stand?

Nanotechnology in the twenty-first century

As an idea, 'nanotechnology' has many sceptics. There are many who claim that 'nano' is a buzzword, that 'nanotechnology' is simply a convenient marketing technique for companies to sell products and for engineers to raise funding. Looking back on the history of the field, it is not difficult to see where that belief comes from. It is certainly true that turn-of-the-millennium nightmares, of nanotechnologists spawning out-of-control nanomachines, have not come to pass.

However, nanotechnology is still in its earliest days. To better understand where nanotechnology as a whole stands, we might not focus on the machines themselves, but on the materials that form them. We do need the materials before we can build nanomachines, after all.

It's telling how important materials are to humankind that historians use materials as a way of dividing time. The vast majority of *Homo sapiens'* time on Earth was spent in the Stone Age, when our antecedents' most sophisticated technology consisted of tools chiselled from rock. Metalworking is only a few millennia old, emerging at different times in different points around the world. Depending on where you lived, you may have experienced some combination of Copper Age, Bronze Age and Iron Age.

The material in each age's name shaped that age's technology and, more importantly, the societies who worked it into that technology. Bronze does not occur naturally, and you can't mine it directly; you must create it by combining copper with tin. The Bronze Age Mediterranean, then, relied on a sprawling network of trade routes to get copper from Cyprus and tin from as far afield as Afghanistan and Cornwall. Around 1200 BCE, Eurasia's trade routes violently broke down. It may not be a coincidence that the bronze-using civilisations of the Near East collapsed at the same time.

Even later eras of history that aren't directly named after materials still hint at the substances behind their flagship technologies. The Steam Age began when civilisations began consuming en masse a fuel – coal – that could produce energy on an unprecedented scale. The Atomic Age was built on elements that allow us to split the atom for energy in even greater quantities. There are some who would name the late twentieth and early twenty-first centuries the 'Silicon Age' for our society's reliance on silicon-based electronics.

As in prior eras, these materials don't just shape technology, but also the society that uses it. Coal mining, nuclear fuel enrichment and semiconductor manufacturing have each influenced the modern world's politics. Simply because we know about a resource does not mean we have mastered its use. Humans have been burning coal since deep in the mists of prehistory; the Romans were already mining coal on a large scale across the empire and using it to heat baths. Chemists first isolated uranium and silicon in 1789 and 1823, respectively, more than a century and a half before the first controlled fission reaction or the first silicon transistor.

Aluminium shows us another example. Aluminium is so common today that you cannot walk five minutes in many major cities without encountering an aluminium recycling bin, but this wasn't always the case. The Danish scientist Hans Christian Ørsted was the first to isolate aluminium in 1824, but isolating aluminium was about all that Ørsted actually managed to do. Aluminium was difficult to separate from its ores, and chemists could only grasp at fragments of the metal. Victorian aluminium was as rare and precious as silver.

This is why, when the US government had the Washington Monument constructed over their capital, the Monument's designers saw fit to crown it with a moon-coloured aluminium pyramid. The obelisk's builders erected the capstone in 1884, just in time for several chemists to independently develop a process for separating aluminium from bauxite, making the metal accessible on an industrial scale. The price of aluminium plummeted almost overnight.

Even then, aluminium needed several more decades to become the utterly mundane material we know today. It was not until the 1950s that the masses could purchase drinks in aluminium cans. Technology simply needed time to catch up with the material's capabilities.

It's impossible to predict the future with any reasonable degree of accuracy, but it's possible to work with the trends we can see. As of this writing, carbon nanotubes have only been around for less than thirty-five years; graphene has only been in laboratories for just over twenty. In human terms, these materials have barely exited their adolescence.

Take graphene. Today, the world simply can't manufacture enough graphene. Graphene factories do exist, but most rely on top-down methods that are essentially more sophisticated versions of the original scotch-tape approach. The ragged graphene that emerges is more than adequate as an additive or as a lubricant, but graphene transistors call for graphene of a far higher quality.

High-end graphene only comes from bottom-up methods like chemical vapour deposition, which are unfortunately still expensive.

Again, it's impossible to predict the future with certainty, but it *is* possible to say with certainty that many researchers are interested in graphene. That fact makes it likely that a method for cost-effectively growing graphene will emerge in a matter of years, rather than decades or centuries.

What, then, might we expect to see from nanotechnology as it continues growing?

The next several years

In short, many of the nanomaterials and nanotools we've discussed are likely to keep developing. Even if these developments play out slowly and unspectacularly, they are still playing out.

We are likely to see many more advances from our ability to manipulate biological machinery. mRNA vaccines carry the momentum from their success against Covid-19. The number of mRNA vaccines in the works rose more than sevenfold between 2019 and 2024. The annual flu vaccine may soon enter our cells on nanoparticles, and mRNA vaccines are now in trials for diseases that have no other vaccine, like Zika and HIV. We're also starting to see trials for personalised mRNA cancer vaccines. Thousands of NHS cancer patients in England may soon get bespoke mRNA shots tailored against their specific cancer cells and the specific mutations lurking within.

Likewise, CRISPR continues to progress. While future progress is never guaranteed, it is promising that researchers needed little more than a decade to turn CRISPR from a lab curiosity to a Nobel-winner to an approved tool that physicians are actively using against long-fearsome disease. Now, early-stage trials are under way for gene treatments that would pit CRISPR against cardiovascular disease, HIV, diabetes and lupus.

In the semiconductor space, silicon's pace towards smaller sizes keeps slowing to a crawl. Even if silicon is unchallenged for now, it seems like a matter of time before 2D semiconductors begin to overtake it. The first successful challengers could come in the form of specialised chips for specific uses; soon, your computer or mobile might connect to Wi-Fi with a graphene-based radiofrequency chip.

We probably won't see silicon electronics completely phase out at once – indeed, the most successful silicon challenger may actually be the one which can cooperate with silicon best, the one that today's silicon-oriented chipmakers can best make with their existing tools. Carbon nanotubes may have an advantage there; in 2020, MIT researchers showed off carbon nanotube transistors they'd fabricated with tools you might find in a typical silicon foundry.

We're also likely to see more and more uses for quantum dots, especially as we get better at making blue dots. We'll of course see more of our televisions and computer screens use quantum dots, and away from public eyes, quantum dots could light up the insides of cancer cells.

Researchers have already shown that these quantum dot tracers (which tend to be made from non-toxic silicon, rather than cadmium) latch onto cancer cells in lab mice and glow fluorescent when imaged. There's also plenty of interest in using quantum dots as pollutant detectors. We can tune them and release them into a vat of wastewater to detect toxins.

Speaking of the environment, we'll likely see more industrial-scale plants scrubbing greenhouse gas with nanomaterials or with filters fashioned from MOFs or COFs. They won't be a panacea for global warming, but they may show that carbon capture has a very real place in the world of tomorrow. On the ground, researchers have shown that thin, flexible solar panels with graphene electrodes are viable and effective. In the lab, these solar panels will undoubtedly keep becoming even more flexible. It's likely they will soon start leaving the lab as engineers find ways of mass-producing them.

We'll surely see more and more nanomaterials find their way into the physical world all around us. Construction materials like smart windows and nanoparticle-enhanced concrete already exist, and we may see more buildings use them. Mass production of graphene-enhanced carbon fibre has now begun, and we may see graphene-enhanced aircraft parts before long.

All signs indicate that we're in nanotechnology's very earliest days. The sort of nanotechnology that is already widespread tends to take advantage of nanomaterials' special properties: the graphene sheet's conductivity, the quantum dot's colour-customisation. The likes of mRNA vaccines and DNA computers are another step, in which we pick up more complex tools already lying around in the nano realm and learn to use them for ourselves. These tools are powerful, and though we've already found plenty of uses for them, we've barely scratched the surface of what they can do.

Molecular moving parts, despite winning a Nobel, are still in their infancy; so are molecular computers. We know these abilities are possible, and we know that scientists are interested. What it will take is research, funding and time. By then, we might find that our space elevator is ready to launch.

FURTHER READING

There's surprisingly little published about the world of the very small. Much of it dates back to the 1990s or 2000s when the word 'nanotechnology' first captured the world's imagination, and newer breakthroughs like mRNA vaccines are simply too recent. Here are several books and online sources that can help.

Chapter 1: Zooming in

Space Elevators: A History: https://www.isec.org/studies/#history – as the name suggests, a detailed history of the concept, from an organisation advocating for the construction of one.

Chapter 2: The world's tiniest assembly guide

Graphene, Brian Clegg (Icon, 2018) – tells the story of graphene's discovery and many of the material's possible applications. Still, graphene science is fast-advancing, and quite a few advances have taken place since its publication.

Chapter 3: Racing to the bottom

'There's Plenty of Room at the Bottom', Richard Feynman (1959): https://web.pa.msu.edu/people/yang/RFeynman_plentySpace.pdf – often credited as the speech that kicked off nanotechnology, though as we've seen, whether Feynman actually started nanotechnology is dubious. What is less dubious is that many nanoscientists have taken Feynman's words as inspiration in the decades since.

Engines of Creation, K. Eric Drexler (Anchor Library of Science, 1986) – we've encountered it more than once in this book. Although controversial, somewhat dated and prone to going on tangents, it's still an interesting look at some people's dreams of what nanotechnology can accomplish. Drexler's ideas of biological assemblers are especially interesting in hindsight.

Chapter 4: A fantastic voyage

The tangled history of mRNA vaccines, Elle Dolgin (2021): https://www.nature.com/articles/d41586-021-02483-w – a great overview of how the technology lived in the background for decades before the Covid-19 pandemic raised them from obscurity.

The 2020 Nobel Prize in Chemistry: https://www.nobelprize. org/prizes/chemistry/2020/summary/ – within the materials for Charpentier and Doudna's prize, you'll find an excellent introduction to CRISPR from a biologist's perspective.

The Code Breaker, Walter Isaacson (Simon & Schuster, 2021) – a more in-depth look at CRISPR via a biography of Jennifer Doudna, one of the key scientists behind its discovery.

Chapter 5: Electrons at the nano gates

The 2023 Nobel Prize in Chemistry: https://www.nobelprize.org/ prizes/chemistry/2023/summary/ – Yekimov, Brus and Bawendi's prize announcement material has more about quantum dots, how they were created and especially how scientists have used them.

Chapter 6: Going small to save the planet

Climate tech news: https://spectrum.ieee.org/topic/ climate-tech/ – a great source for the latest research in energy technology, including solar panels, batteries and more. Much of it features nanomaterials.

Omar Yaghi's Ullyot Public Affairs Lecture: https://www. sciencehistory.org/about/awards-program/ullyot-public-affair s-lecture-and-award/ – a more in-depth look at MOFs and COFs, and their environmental applications, from the scientist who helped create them.

Chapter 7: Nano nightmares

Ultrafine particles/nanoparticles, World Health Organization: https://www.who.int/teams/environment-climate-chang e-and-health/air-quality-and-health/videos/mosaic/ ultrafine-particles-nanoparticles – an enlightening video series covering where nanoparticles in the environment come from, how they can harm human health and how scientists study them.

Chapter 8: Nano dreams

The 2016 Nobel Prize in Chemistry: https://www.nobelprize. org/prizes/chemistry/2016/summary/ – complete with more examples of the molecular machines that won Sauvage, Stoddart and Feringa the award.

Nanocar Race II: https://www.memo-project.eu/flatCMS/ index.php/Nanocar-Race-II – there you can find details on the contestants and YouTube links to the entire race, if you have the wherewithal to watch a 24-long race.

Plastic Fantastic: How the Biggest Fraud in Physics Shook the Scientific World, Eugenie Samuel Reich (Macmillan, 2009) – a detailed account of the Jan Hendrik Schön affair. While it is primarily the telling of a scientific scandal (and a good read if that interests you), it's also a look into the nanoscience at hand.

Chapter 9: Zooming out

Nano Supermarket: https://www.nanosupermarket.org/ products – a stylised look at the sorts of consumer products that might contain nanomaterials in just a few years' time.

The Kavli Prize in Nanoscience: https://www.kavliprize.org/ prizes/nanoscience/2024 – as far as cutting-edge nanoscience is concerned, the Kavli Prizes are perhaps second only to the Nobels in prestige. The work that wins each year is a good indicator of where the field is headed.

ACKNOWLEDGEMENTS

I wish to specially thank Himani Galagali at Massachusetts General Hospital, Mark Hersam at Northwestern University and Mark Wiesner at Duke University for generously providing their expertise during this book's creation. Additional thanks to Judy Meiksin, Lin Liu, Cooper Garrison, Brian Clegg and the team at Icon Books for their help behind the scenes. This book would not have been possible without their assistance and contributions.

INDEX

Adleman, Leonard M. 158–9
aerogel 109–10
aluminium 169–70
aluminium formate 114
Alzheimer's disease 64
Analytical Engine 85n
aquaporins 118
arsenic 120
asbestos 133
atoms 14–15
Aviram, Arieh 153–4

Babbage, Charles 85n
bacteriophages 70
barium meal examination 59
batteries 106–8
Bawendi, Moungi G. 76,
  80, 83
Bell Labs 83–4, 154–5, 163
Binning, Gerd 43
Bohr, Niels 15
breaking length 3
British Museum 35–7
Bronze Age 168

Brus, Louis E. 76,
  79–80, 83
buckminsterfullerene 48–9,
  156 *see also* fullerenes
buckyball *see*
  buckminsterfullerene

CALF-20 114
cancer
  glioblastoma 62, 64
capacitors 108
carbon
  atom 15
  capture 111–12, 114–16,
    173
  chains 46–7
  diamond 16–18
  electronics 88–91
  graphene 13–14, 27–32,
    43n, 90–1, 104–5, 107,
    108, 115–16, 117, 155,
    170–1, 172, 173
  graphite 13, 16–18, 20,
    28–30, 47, 107

nanotubes 3–4, 8, 9–11, 17, 20, 30–1, 33, 37, 49, 53, 62, 88–90, 118, 119, 133, 170, 172
Cas9 70–2
catenanes 147–8
Charpentier, Emmanuelle 70–2, 73
Chavez, Hugo 128
chemical vapour deposition (CVD) 32–3
Clarke, Arthur C. 1
COFs 114–15, 173
computers
    DNA-based 157–60
    molecular 153–6
    spintronic memory 97–8
    ternary 157n
concrete 110–11
covalent bonds 115, 154
covalent organic frameworks see COFs
Crichton, Michael
    Prey 124
CRISPR 69–73, 172
crystal structure 17
Curl, Robert 47, 49
CVD *see* chemical vapour deposition

Da Jiang 28
Dalton, John 14–15, 88
Damascus steel 37
de Broglie, Louis 77
de Broglie wavelengths 77, 78
Democritus 14
desalination 117–18
diamond *see* carbon
dichroic glass 36–7
DNA 66, 69, 70, 157–63
Doudna, Jennifer 70–2, 73

Doxil 52–3, 62
Drexler, Kim Eric 50–2, 54, 55, 150–1
    Engines of Creation 51–2, 125–6

E. coli 69, 124
Eigler, Donald M. 44–5
electrons 15, 18
elevator, space 1–4, 9–10, 53, 164–6
Engines of Creation (Drexler) 51–2, 125–6

Fantastic Voyage 57–8, 146
Ferina, Ben 149, 152
fertilisers 136–7
Feynman, Richard 8, 38–9, 146, 150, 163
Fuller, Buckminster 45–6
    see also buckminsterfullerene; fullerenes
fullerenes 49, 104, 105n *see also* buckminsterfullerene

geckos 22–3
Geim, Andre 13–14, 27–30, 31–2, 54
glioblastoma *see* cancer
gold 25, 35, 60–1, 79n, 104, 132, 144
graphene *see* carbon
graphene oxide 117–18
graphite *see* carbon

Hamiltonian path problem 158–9
h-BN 95–7
hexagonal boron nitride *see* h-BN
Hibbs, Albert 146

HIV 63
human immunodeficiency
    virus *see* HIV

Iijima, Sumio 49
Institute for Soldier
    Nanotechnologies *see* ISN
insulation 109–10
International Space Station 2
ionic bonds 114–15
ISN (Institute for Soldier
    Nanotechnologies) 127

Joy, Bill
    'Why the Future Doesn't
        Need Us' 54n

Kroto, Sir Harry 45–9

LEDs 83 *see also* QLED
    displays
Leucippus 14
Lieber, Charles 128–9
liposomes 63, 67–9
lotus 23–4
lubricants 111
lustreware 36
Lycurgus Cup 35, 79n

machines, molecular 147–52
magnetic resonance imaging
    see MRI
magnetite 119–20
mesocosms 140
Metal Gear Solid 124
metal-organic frameworks *see*
    MOFs
methane 115–16
microscopes
    electron 41–2
    light-based 41

scanning tunnelling (STM)
    42–5
molybdenum disulphide
    see MoS2
Moore's law 86
MOFs 112–15, 117, 117, 173
MoS2 92–3, 108
motor, molecular 149, 152
MRI 61, 137
mRNA 66–9, 171
MXenes 107–8

nanobots 65, 119–21,
    123, 143
nanofiltration 117
nanoparticles 24–5
    sizing 139
nanostructures 25–7
nanotoxicology 130–3,
    138–9
nanotubes *see* carbon;
    nanostructures
nanoscalpels 62, 64–5
'New Breed, The', *see Outer
    Limits, The*
Nobel prize nomination
    process 76–7
No Time to Die 123–4
Novoselov, Konstantin 13–14,
    27, 29–30, 31–2, 54

oil spills 118–19
O'Neill, Gerard 50n
orbit, geostationary 2
organelles 7, 65
Ørsted, Hans Christian 169
osmosis, reverse 117
Outer Limits, The 124

Palazzo Italia 116
Parkinson's disease 64

particulate matter (PM)
  emissions 134–7
Penzias, Arno 83n
periodic table 88
perovskites 105–6
phosphorus, black 93–4
photolithography 87
platinum 26
PM emissions *see* particulate
  matter
pollution 134–7
polyalkylcyanoacrylate
  129–30
polymers 39–40
Prey (Crichton) 124

QLED displays 80
quantum computing 98–9
quantum dots 76–83,
  99–100, 102–3, 131,
  172–3
quantum mechanics 21
  spin 97
  uncertainty principle 21
  wave-particle duality 21

Ratner, Mark 154
red giants 46
RGD peptide 60–1
ribosome 66, 156–7
RNA 71, 72 *see also* mRNA,
  siRNA
robots *see* nanobots
rockets 1
Rohrer, Heinrich 43
Röntgen, Wilhelm 59
rotaxanes 148–9
Royal Swedish Academy of
  Sciences 75–6
Rutherford, Ernest 15

Sauvage, Jean-Pierre
  147–8, 152
scanning tunnelling
  microscope *see* microscopes
Schön, Jan Hendrik 155
Schweizer, Erhard K. 44–5
Seeman, Nadrian 161
semiconductors 82–4, 89–96,
  101, 102, 172
sewage treatment 118
Shockley–Queisser limit 102
silicon 84–7
  silicene 94–5
silver 35, 36, 104, 118,
  131–2, 138
siRNA 68
Smalley, Richard 47, 49, 151
solar cells 101–6, 173
solar sail 50
STARS-EC 165–6
STM *see* microscopes
Stoddart, Fraser 148, 152
supercapacitors *see* capacitors
surface plasmon resonance 79n

Taniguchi, Norio 40–1
titanium dioxide 25, 52, 110,
  116, 132, 138
TMDs 92–3
topological chemistry 148
transistors 84–7, 88–91, 93,
  94, 97, 155
transition metal
  dichalcogenides *see* TMDs
travelling salesman problem
  see Hamiltonian path
  problem
Tsiolkovsky, Konstantin 164

ultrasound 61, 144

vaccines 67–9
van der Waals, Johannes
    Diderick 19, 20, 21
van der Waals forces 19–20,
    23, 67, 92
viruses 66–7

water 116–18
weapons 126–9
'Why the Future Doesn't
    Need Us' (Joy) 54n

Wilson, Robert 83n
windows, smart 109

X-rays 59–61, 87

Yaghi, Omar 112
Yekimov, Aleksey 76, 78, 79,
    80, 83

zeolites 112n